Designing Delay-Tolerant Applications for Store-and-Forward Networks

For a listing of recent titles in the
Artech House Space Applications Series,
turn to the back of this book.

Designing Delay-Tolerant Applications for Store-and-Forward Networks

Edward J. Birrane
Jason A. Soloff

ARTECH HOUSE
BOSTON | LONDON
artechhouse.com

Library of Congress Cataloging-in-Publication Data
A catalog record for this book is available from the U.S. Library of Congress.

British Library Cataloguing in Publication Data
A catalog record for this book is available from the British Library.

ISBN-13: 978-1-63081-628-5

Cover design by John Gomes

© 2020 Artech House
685 Canton Street
Norwood, MA 02062

All rights reserved. Printed and bound in the United States of America. No part of this book may be reproduced or utilized in any form or by any means, electronic or mechanical, including photocopying, recording, or by any information storage and retrieval system, without permission in writing from the publisher.

All terms mentioned in this book that are known to be trademarks or service marks have been appropriately capitalized. Artech House cannot attest to the accuracy of this information. Use of a term in this book should not be regarded as affecting the validity of any trademark or service mark.

10 9 8 7 6 5 4 3 2 1

This book is dedicated to the life and memory of Adrian Hooke, whose inexhaustible passions for both international collaboration and common sense led to the establishment of some of the institutions, technologies, and aspirations discussed in these pages

Adrian Hooke's work led to the creation of the Consultative Committee for Space Data Systems. His technical insights and persistent advocacy were foundational to establishing the space delay-tolerant networking community and the concepts and structure of the Solar System Internet. His humor, kindness, and camaraderie are missed by those privileged enough to have worked with him on our most astronomical problems.

Adrian is pictured here receiving a lifetime achievement award, inscribed as follows:

The Solar System Internetwork (SSI)
Sculpture by Terrie Bennett
Presented April 2012 to
Adrian Hooke
From his grateful colleagues,
In recognition of his lifelong leadership in CCSDS

Contents

CHAPTER 1

Introduction 1
1.1 The State of the Wireless World 1
1.2 Why Be Patient in an Increasingly Connected World? 3
1.3 What Is a Delay-Tolerant Application? 4
1.4 Who Should Read This Book? 7
1.5 How to Use This Book 8
1.6 Summary 9
1.7 Problems 9
References 10

CHAPTER 2

A Brief History of Challenged Networking Environments 13
2.1 What is a Challenged Networking Environment? 13
    2.1.1 Separating Responsibilities 14
    2.1.2 Defining Network Service Layers 15
2.2 Link Layer Challenges 16
    2.2.1 High-Rate Wireless Communications 16
    2.2.2 Node Mobility 18
    2.2.3 Link Heterogeneity 21
2.3 Network Layer Challenges 22
    2.3.1 Path Losses 22
    2.3.2 Time-Variant, Partitioning Topologies 23
    2.3.3 Unsynchronized Node Information 23
2.4 Application Layer Challenges 24
    2.4.1 Growing Delays 24
    2.4.2 Increased Disruptions 25
    2.4.3 Growing Data Volumes 25
2.5 Error Handling in Challenged Networking Environments 26
2.6 What Is a Network Error Condition? 26
2.7 Approaches to Handling Error Conditions 27
    2.7.1 Increase the Number of Nodes 27
    2.7.2 Alter Link Characteristics 29
    2.7.3 Increase Protocol Efficiency 29

2.8	Summary	30
2.9	Problems	31
	References	31

CHAPTER 3
How the Internet Does It: Approaches and Patterns for Challenged Networking Environments — 33

3.1	Challenges in the Terrestrial Internet	33
	3.1.1 Network Policies	33
	3.1.2 Data Workflows	34
	3.1.3 Network Pricing and Economics	35
	3.1.4 Application Behavior	36
3.2	Terrestrial Internet Approaches to Challenged Networking Environments	36
	3.2.1 Place Information	37
	3.2.2 Push Data	39
	3.2.3 Avoid Sessions	40
3.3	Terrestrial Internet Design Patterns	43
	3.3.1 Content Delivery Networks	43
	3.3.2 Data Subscriptions	45
	3.3.3 Autonomic Computing	47
	3.3.4 Stateless Data	50
3.4	Summary	51
3.5	Problems	52
	References	52

CHAPTER 4
Rallying the Research Community: DARPA, NASA, and Disruption Tolerance — 55

4.1	History of Delay-/Disruption-Tolerant Research	55
4.2	NASA and DARPA	56
4.3	International Space Agencies	58
	4.3.1 History of IOP Activities	59
	4.3.2 Establishment of the IOAG	59
	4.3.3 The Space Communications Architecture Working Group	60
	4.3.4 Space Internetworking Strategy Group	64
4.4	IOP Meets the Consultative Committee for Space Data Systems	64
4.5	DTN in the IRTF	65
4.6	Ongoing Development	65
4.7	Summary	66
4.8	Problems	66
	References	67

CHAPTER 5
Where the Terrestrial Internet Is Not Enough: Motivating Use Cases — 69

5.1	The Value of Use Cases	69
5.2	The Solar System Internet	69

		5.2.1	A Brief History of Space Communication	70
		5.2.2	A Solar System Internet	74
		5.2.3	An Architecture for the SSI	78
	5.3	Distributed Spacecraft Constellations		79
		5.3.1	Planetary Observation Missions and Space-Ground Integration	80
		5.3.2	Deep-Space Instruments	82
		5.3.3	In-Space Communication and Navigation	84
	5.4	Distributed and Mobile Sensor Webs		86
	5.5	Optical Communications		87
	5.6	Ad Hoc Network and Data Mules		89
	5.7	Summary		89
	5.8	Problems		89
		References		91

CHAPTER 6

The Delay-/Disruption-Tolerant Networking Architecture 93

6.1	Motivations for a Tolerant Network		93
6.2	Assumptions Made by the Terrestrial Internet		93
	6.2.1	Path Existence	94
	6.2.2	Timely, Reliable, Actionable Feedback	94
	6.2.3	Small End-to-End Data Loss	95
	6.2.4	TCP/IP Ubiquity	95
	6.2.5	Performance Abstraction	95
6.3	Architectures for DTNs		96
6.4	Delay-/Disruption-Tolerant Desirable Properties		96
6.5	DTN Protocols		97
6.6	Naming and Addressing		99
6.7	The BP Ecosystem		99
	6.7.1	Convergence Layers	100
	6.7.2	Convergence Layer Adapters	100
	6.7.3	The BPA	101
	6.7.4	Application Agent	101
6.8	Special Node Characteristics		102
	6.8.1	Persistent Storage	103
	6.8.2	Late Binding	103
	6.8.3	Multiple Convergence Layers	104
6.9	Summary		105
6.10	Problems		106
	References		106

CHAPTER 7

Patience on the Wire: The DTN BP 109

7.1	Protocol Goals	109
7.2	The Case for BP Store and Forward	110
7.3	Services Unique to BP	111
7.4	Protocol Layering Considerations	111

		7.4.1	Versions of the BP	112
7.5	Bundle Structure			113
7.6	The Primary Block			114
		7.6.1	Version	114
		7.6.2	Processing Control Flags	114
		7.6.3	Cyclic Redundancy Check Type	115
		7.6.4	Destination EID	115
		7.6.5	Source Node ID	115
		7.6.6	Report-To EID	115
		7.6.7	Creation Timestamp	115
		7.6.8	Lifetime	116
		7.6.9	Fragment Offset	116
		7.6.10	CRC Field	116
7.7	The Payload Block			116
7.8	Extension Blocks			116
7.9	BP-Enabled Concepts			118
		7.9.1	Application Annotations	118
		7.9.2	Custody Transfer	120
		7.9.3	Content Caching	121
7.10	Special Considerations			121
		7.10.1	Storage Management	121
		7.10.2	Security	122
		7.10.3	Fragmentation	123
		7.10.4	Optimal Fragment Size	123
		7.10.5	Handling Extension Blocks	123
		7.10.6	Additional Processing	124
7.11	Is BP Enough?			125
7.12	Summary			126
7.13	Problems			126
	References			127

CHAPTER 8

Advanced Networking Architectures   129

8.1	Networking Architectures		129
8.2	A Standard Model for Networking		130
8.3	Overlay Networks		131
	8.3.1	Pass-Through and Encapsulating Interfaces	132
	8.3.2	Differing Network Addressing Schemes	134
8.4	Partitioned Networks		139
8.5	Federated Internetworks		140
8.6	Summary		141
8.7	Problems		141
	References		142

CHAPTER 9

Application Services and Design Patterns   145

9.1	A Multitiered Application Service Hierarchy		145
	9.1.1 Transport Services		148
	9.1.2 Core Networking Services		149
	9.1.3 Federating Services		150
	9.1.4 Overlay Services		151
	9.1.5 Endpoint Services		151
9.2	Application Design Patterns		152
	9.2.1 The History and Concept of Design Patterns		153
	9.2.2 The Value of Patterns in Emerging Application Domains		154
9.3	The Design Pattern Documentation Format		156
9.4	Summary		158
9.5	Problems		159
	References		160

CHAPTER 10

The Offshore Oracle Pattern: Caching Content in Challenged Networks   161

10.1	Pattern Context	161
10.2	The Problem Being Solved	165
10.3	Pattern Overview	166
	10.3.1 Role Definitions	166
	10.3.2 Control and Data Flows	168
10.4	Service Types	170
10.5	When and How to Integrate	171
	10.5.1 When to Use This Pattern	171
	10.5.2 Recommended Design Decisions	172
10.6	What Can Go Wrong	173
10.7	Pros and Cons	174
10.8	Case Studies	174
	10.8.1 BP Status Reporting	174
	10.8.2 Topology Management	176
	10.8.3 Security Policy Updates	177
10.9	Summary	178
10.10	Problems	178
	References	179

CHAPTER 11

The Training Wheels Pattern: Open-Loop Control   181

11.1	Pattern Context	181
11.2	The Problem Being Solved	184
11.3	Pattern Overview	185
	11.3.1 Role Definitions	186
	11.3.2 Control and Data Flows	188
11.4	Service Types	190
11.5	When and How to Integrate	192
	11.5.1 When to Use This Pattern	192
	11.5.2 Recommended Design Decisions	193

11.6	What Can Go Wrong	194
11.7	Pros and Cons	195
11.8	Case Study	195
	11.8.1 Spacecraft Fault Protection	195
11.9	Summary	196
11.10	Problems	198
	References	199

CHAPTER 12

The Stow Away Pattern: Annotated Messaging — 201

12.1	Pattern Context	201
12.2	The Problem Being Solved	204
12.3	Pattern Overview	206
	12.3.1 Role Definitions	206
	12.3.2 Control and Data Flows	208
12.4	Service Types	208
12.5	When and How to Integrate	209
	12.5.1 When to Use This Pattern	209
	12.5.2 Recommended Design Decisions	211
12.6	What Can Go Wrong	212
12.7	Pros and Cons	213
12.8	Case Studies	213
	12.8.1 The BP Security Extensions	214
	12.8.2 Contact Graph Routing Extensions	215
12.9	Summary	216
12.10	Problems	217
	References	217

CHAPTER 13

The Network Watchdog Pattern: Distributed Error Detection and Recovery — 219

13.1	Pattern Context	219
13.2	The Problem Being Solved	223
13.3	Pattern Overview	224
	13.3.1 Role Definitions	226
	13.3.2 Control and Data Flows	228
13.4	Service Types	229
13.5	When and How to Integrate	231
	13.5.1 When to Use This Pattern	231
	13.5.2 Recommended Design Decisions	232
13.6	What Can Go Wrong	233
13.7	Pros and Cons	234
13.8	Case Studies	234
	13.8.1 BP Administrative Records	234
13.9	Summary	236
13.10	Problems	237
	References	238

CHAPTER 14

The Data Forge Pattern: Leveraging In-Network Storage — 239

- 14.1 Pattern Context — 239
- 14.2 The Problem Being Solved — 242
- 14.3 Pattern Overview — 243
  - 14.3.1 Role Definitions — 244
  - 14.3.2 Control and Data Flows — 245
- 14.4 Service Types — 247
- 14.5 When and How to Integrate — 247
  - 14.5.1 When to Use This Pattern — 247
  - 14.5.2 Recommended Design Decisions — 248
- 14.6 What Can Go Wrong — 250
- 14.7 Pros and Cons — 250
- 14.8 Case Studies — 251
  - 14.8.1 Store-and-Forward Routing Applications — 251
  - 14.8.2 Information-Centric Networking — 252
- 14.9 Summary — 253
- 14.10 Problems — 253
- References — 254

CHAPTER 15

The Ticket to Ride Pattern: Regional Administration — 257

- 15.1 Pattern Context — 257
- 15.2 The Problem Being Solved — 259
- 15.3 Pattern Overview — 262
  - 15.3.1 Role Definitions — 263
  - 15.3.2 Control and Data Flows — 263
- 15.4 Service Types — 265
- 15.5 When and How to Integrate — 265
  - 15.5.1 When to Use This Pattern — 265
  - 15.5.2 Recommended Design Decisions — 266
- 15.6 What Can Go Wrong — 267
- 15.7 Pros and Cons — 268
- 15.8 Case Studies — 268
  - 15.8.1 Federated Deep Space Networking — 268
- 15.9 Summary — 270
- 15.10 Problems — 271
- References — 271

CHAPTER 16

What Can Go Wrong Along the Way: Special Considerations for DTNs — 273

- 16.1 Resource Limitations — 273
- 16.2 Accepting Reactive Fragments — 275
- 16.3 Heterogenous Networks — 275
- 16.4 Dissimilar Implementations — 276

		16.4.1 Dissimilar BPA Extensions	277

        16.4.1 Dissimilar BPA Extensions 277
        16.4.2 Dissimilar Service Level Expectations 277
        16.4.3 Dissimilar Operating Environments 278
16.5 Noah's Data Ark: A Case Study 280
16.6 Working within an Overlay Network 281
        16.6.1 Routing and Name Spaces 281
        16.6.2 Networks in Motion 282
        16.6.3 What's in a Name? 283
16.7 A Network Is a Network 283
16.8 Summary 283
16.9 Problems 284
      References 284

CHAPTER 17

The Solar System Internet: A Case Study for Delay-Tolerant Applications 287

17.1 Overview 287
17.2 Motivation 288
17.3 Experiences and Experiments 289
        17.3.1 Mars Relay Communications 289
        17.3.2 Deep Impact Networking Experiments 290
        17.3.3 Multi-Purpose End-To-End Robotic Operation Network 292
        17.3.4 International Space Station 294
17.4 Significant Challenges 295
        17.4.1 Connectivity 296
        17.4.2 Delay 296
        17.4.3 Bandwidth 296
17.5 Similarity to Emerging Challenged Terrestrial Networks 297
        17.5.1 Remote Sensing Through IoT Devices 297
        17.5.2 Wildlife Tracking 298
        17.5.3 Vehicle Data Logging 299
        17.5.4 Search and Rescue Networks 299
17.6 Features Enabled by Application Patterns 300
        17.6.1 Autonomous Network Management 300
        17.6.2 Inter-Domain Communications 302
        17.6.3 Coordinated Data Fusion 304
17.7 Summary 305
17.8 Problems 306
      References 307

About the Authors 309

Index 311

CHAPTER 1
Introduction

This book provides an overview of delay-tolerant networking (DTN) [1] and defines several patterns for the design of applications operating in challenged networking environments. This information is generally useful to anyone learning about networking computers in exotic or otherwise nontraditional ways and specifically focused on how this can be accomplished using transport protocols specific to the DTN family of networking solutions. An important observation that motivates this work is that networking solutions such as DTN enable new networking architectures but also require a new approach to network application programming.

The operational deployment of DTNs cannot be transparent to networking (or user) applications if these applications have performance requirements that are not themselves delay-tolerant. When delays are likely, applications must be able to accept them as part of the normal functioning of the network and not as a transient error condition followed by some future recovery to a nondelayed state. This work discusses the ways in which DTNs differ from the terrestrial internet, why they are growing in popularity, and how they require new thinking from networking architects, application programmers, and users/operators.

1.1 The State of the Wireless World

Computer internetworking research and development focuses on maintaining the assumptions, algorithms, and capabilities of the terrestrial internet as the number of users and the amount of user data increases. Because most such research is focused on incremental efficiencies to maintain the user experience, a common set of metrics can be defined and used to determine how (and how much) a new technique maintains capability against an increasing digital workload.

Marketing metrics such as upload and download speeds can be deceptive in describing the overall user experience in a network; a very fast download speed can be offset by very inefficient use of networking resources. A better indicator of network (or algorithm) performance is one that speaks to efficiency. Two common efficiency metrics for computer networking are called goodput and throughput. Throughput is the overall amount of data that must be sent to communicate some set of user information. This includes the user data itself, but also protocol overhead, security, control-plane signaling, and retransmissions. In the context of commercial networks, activities such as in-service upgrades, network management, and security processing also count toward throughput.

Goodput is the subset of throughput (see Figure 1.1) representing the user data itself. The efficiency of the network can then be defined by the ratio: efficiency = goodput/throughput. For example, if it takes 100 bytes to send 100 bytes of user data, the network is perfectly efficient with a ratio of 1.0. If it takes 200 bytes to send 100 bytes of user data, the network is half as efficient, with a ratio of 0.5. Higher efficiencies are preferred because a more efficient network spends less time transmitting nonuser data and the overall latency experienced by the user at endpoints in the network is decreased.

The popularity of efficiency and latency as important metrics bias how networks evolve. The innovations pursued by academia and industry will be prioritized by what impacts those metrics because they are the commonly accepted definition of good. Enterprise solutions focus on increasing resources: more cellular towers, more wireless routers, more computers in the data center. Research approaches seek to optimize existing data representations: more symbols, faster switching, reduced overhead, better compression. These approaches usefully optimize existing infrastructure but fail in challenging scenarios where infrastructure investment is impossible or impractical.

For example, 4G LTE cellular providers started to discontinue unlimited data plans as user data volumes increased (e.g., high-definition video) and the number of users increased (e.g., the ubiquity of smartphones) [2]. The extension of 4G assets to handle this scenario was deemed impractical and new mechanisms to include

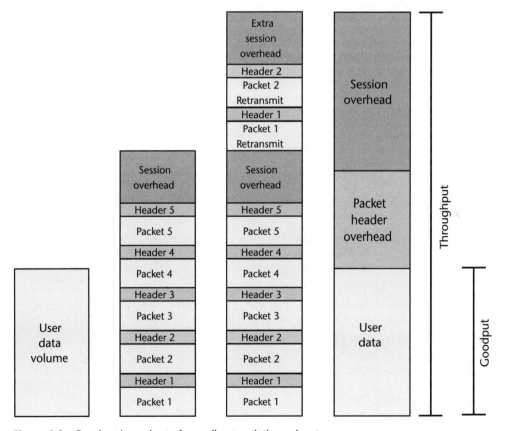

Figure 1.1 Goodput is a subset of overall network throughput.

5G service are being developed to handle this situation instead. Wireless routers in crowded conference hotels drop connections when handling the surge of thousands of users. Maintaining surge infrastructure is not cost effective for transient demands. Similarly, maintaining constant infrastructure in unpopulated service areas to include oceans, deserts, and space is impractical.

This is not a criticism of our current approach to networking. Terrestrial network engineering companies rightfully focus on increasing the throughput of already high-speed, ubiquitous wireless services in densely populated parts of the world. Their envisioned next-generation systems (such as 5G cellular) offer more of the same: faster data rates in high coverage areas to meet increasing user data needs. These optimizations, while necessary, do not address the different and growing need for data connectivity in increasingly remote areas.

1.2 Why Be Patient in an Increasingly Connected World?

"If I had more time, I would have written a shorter letter." Blaise Pascal expressed this sentiment in a collection of letters—Lettres Provinciales—published in 1657. This saying is a popular one because it captures an important truth: We regularly sacrifice efficiency for expediency.

When applied to networking, expediency refers to the ability to rapidly disposition a user message (e.g., a packet) with minimal processing by the source of the message and the nodes that forward it along its path. If a terrestrial network provider must choose between making the network more efficient or making the network appear faster to a customer, it will usually invest in the customer experience. Why spend resources refactoring and optimizing when purchasing faster computers or installing more points of presence will make the system go faster without any rework? This approach, often referred to as throwing hardware at the problem, does not scale indefinitely.

Growing a network by purchasing and deploying more and more capable hardware is, at best, an expensive and hard to maintain strategy as the volume of user data outpaces the ability to deploy networking infrastructure. The strategy becomes impossible when connecting remote parts of the world or the solar system [3]. When it becomes impossible or impractical to adapt infrastructure to data requirements, then data requirements must adapt to the limitations of the infrastructure. Adapting to network limitations, in this context, means analyzing what information exchange must occur versus what information exchange exists only as an artifact from inefficient reuse of tools, protocols, or algorithms from less constrained environments.

Recognizing that while we live in a real-time world, much of our data is not real time, a group of network protocol engineers proposed the concept of a DTN [1, 4, 5]. DTNs safely and securely communicate information along a networking path that cannot otherwise accommodate expedient data delivery. Because of their ability to deliver data in challenging networking conditions, they offer a pragmatic alternative to the concept of ubiquitous, high-rate wireless networks; they are required for internet-style communications in situations where near-instantaneous end-to-end connectivity is impractical or impossible.

DTN, as an architectural approach, standardizes how nodes persist data while waiting for an opportune time to transmit it to its next networking hop. This is a conceptually simple, low-level, but fundamentally different approach to networking than that taken on in the terrestrial internet, where a message must either be forwarded, delivered, or abandoned within a very small timeframe (typically milliseconds). The most significant protocol supporting the development of DTNs is the Bundle Protocol (BP) [6], which provides the packet structure necessary to implement information storage and describes the behavior of BP agents in storing and otherwise managing these packets.

The architectures, features, and protocols used to implement DTNs are covered in much greater detail in subsequent chapters. However, there are several simple examples of why storing a message at a waypoint node in a network may be beneficial. Figure 1.2 illustrates two approaches to routing a message in a network, from Node 1 to Node 4 (shown on the vertical axis) of a series of messages in a dataset, D1. The first approach, Figure 1.2(a), is typical of a terrestrial style of message delivery. In this approach, the messages comprising D1 wait until an end-to-end link exists in a network, resulting in its expeditious delivery relative to the time at which messages in D1 were transmitted. However, the communication of these messages so congests the network that some other dataset, D2, is unable to be communicated. The second approach, Figure 1.2(b), uses a different style of message delivery where messages in D1 are communicated from the source as soon as any network link is available, even if that link does not represent a direct path to the dataset destination. In this case, the messages in D1 arrive at their destination faster while also allowing messages from dataset D2 to also be delivered.

The strategy of storing data in the network is not needed in highly available, never-congested networks because messages in those networks are never delayed long enough to require being stored. However, as delays (and disruptions) are added to the networking environment, new approaches to data exchange need to be reviewed. When faced with congestion, latency, and other types of impairments, storing data in the network results in faster delivery of more data (by not having to repeatedly retransmit) and therefore a higher data efficiency.

1.3 What Is a Delay-Tolerant Application?

A delay-tolerant application is one that continues to work without requiring a fast, reliable communications infrastructure; the application tolerates delays in message delivery (or acknowledgment delivery) because of impairments in the network itself. Applications that connect to high-speed, reliable networks today—such as those operating over the terrestrial internet—are dependent on continued infrastructure investments to prevent delays. When these networks experience delays due to congestion, poor signal reception, natural disaster, or other impairments delay-intolerant applications stop working. Worse, if such intolerant applications cannot degrade gracefully, they may flood the local network with attempts to reestablish connectivity. Such floods can prevent network recovery, consume local node resources, and effect the operation of coresident applications.

The most natural milieu for the deployment of a delay-tolerant application is a DTN. DTNs support end-to-end communications without requiring an

1.3 What Is a Delay-Tolerant Application?

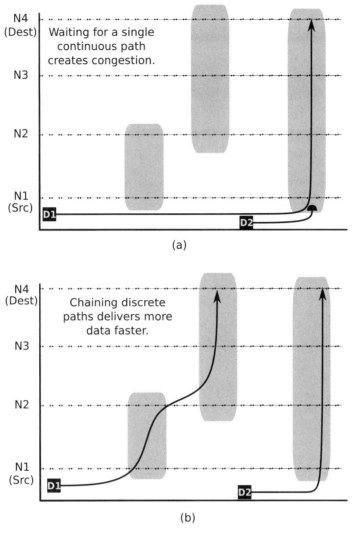

Figure 1.2 Delay tolerance trades efficiency for expediency.

instantaneous, end-to-end path by introducing new packet structures and new messaging semantics that eliminate dependencies on round-trip data exchange. There has been significant research and engineering effort related to understanding how common networking services must be reimagined around store-and-forward techniques. For example, DTN waypoints can store individual messages for long periods of time [7, 8]; individual DTN packets can carry and accumulate annotative, session-like information; DTN protocols can enable intelligent reporting, timing, and retransmission logic [9]. These unique features allow the network to tolerate the myriad of delays a packet may encounter in a challenged networking scenario. Delays caused by long signal propagation times, intermittent pauses due to loss of an individual link, and queuing caused by node-local administrative/prioritization policies are all better handled by a DTN network than an Internet Protocol (IP) network [10–12].

Reviewing the ways in which the capabilities, assumptions, and constraints of a DTN differ from the terrestrial internet illuminates the ways in which application engineering must be thought of differently. However, networks exist so that applications can exchange data over them and those applications must also be designed to survive in a challenged networking scenario; the tolerance of the network does not necessarily bestow tolerance on the application. An application built in an intolerant way will not work even when operating over a DTN. For example, applications such as traceroute that assume round-trip communications faster than network topological change do not produce useful results. Figure 1.3 illustrates an example flow of data through a DTN.

In this figure, we assume a non-delay-tolerant application (sender) and a non-delay-tolerant application (destination) that are using a DTN to exchange messages (Msg M1, M2) and their respective acknowledgments (ACK A1, A2). When M1 is sent to the network, it is stored-and-forwarded at the intermediate nodes between the sender and the destination. However, if the sender is unaware of the possibilities of message delay, it may implement its own timeouts/resets resulting in a canceling of M1 and the issuing of a duplicate message, M2. In cases where

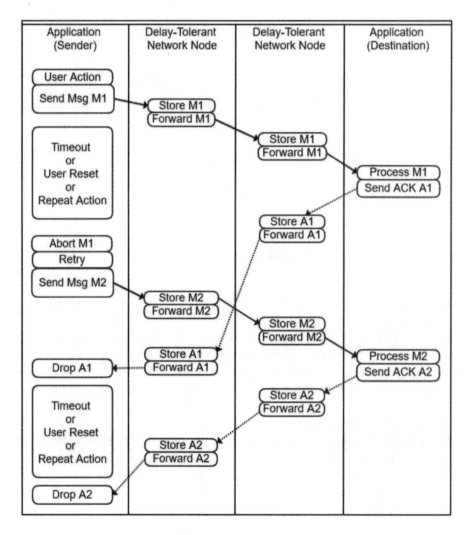

Figure 1.3 DTNs do not guarantee delay-tolerant applications.

the delays in the network are longer than the timeouts configured in the sending application, messages may never be acknowledged because the timeout before the acknowledgments can be received by the sender. This is the situation shown in Figure 1.3. In some cases, the regeneration of a message may have no impact beyond creating additional networking traffic. In other cases, regeneration of a message may involve aborting higher-level transactions, renegotiating session information, requiring users to re-assert some credentials, or incurring other system-impacting consequences. When a network provides delay tolerance, networking applications must be written to rely on those networking mechanisms and not try to re-create their own. Just as common internet applications operating over Transmission Control Protocol (TCP) do not need to implement their own fragmentation and reassembly scheme, common DTN applications should not try to implement their own store-and-forward mechanisms.

1.4 Who Should Read This Book?

This book is intended as a resource for systems engineers, network architects, application developers, and network operators when answering the question: "We are deploying a DTN, but what do we actually do with it?"

System engineers develop the concept of operations (CONOPs) for missions that use networking resources. They develop the concepts of data volumes and data flows, derive performance requirements, and make design decisions based on networking constraints and infrastructure assumptions. Delay tolerance is, fundamentally, a system design decision and this book provides systems engineers with the information necessary to understand how this approach changes networking CONOPS used to operate in more constrained environments. Certain assumptions (such as high availability, high reliability, end-to-end paths) are relaxed while other assumptions (in-network storage) are added. Adopting a delay-tolerant approach to system design provides a novel way to migrate familiar CONOPs to networks that cannot otherwise provide common internet-style guarantees.

Network architects and engineers design and implement the logical and physical components of individual networking builds. This may include the routing and protocol configuration for overlay networks, the provisioning and integration of hardware comprising an intra-network, or the construction of highly specialized equipment for unique cases. Network engineers define the required link services for a network; delay tolerance in networking applications informs hardware and link configuration and provisioning. As such, this book describes a useful alternative to hardware provisioning when extending networking concepts into resource-challenged areas. Understanding how to incorporate delay tolerance into a network may dramatically reduce the cost and maintenance associated with network architecture and deployment.

Application developers have the challenging task of designing and implementing applications that both run over the infrastructure provided by network architects, within the constraints provided by network engineers, and conforming to the concepts and functionality specified by systems engineers. This book provides specific recommendations, design considerations, and motivations for how to make applications work in challenged networks. Achieving logical functionality without

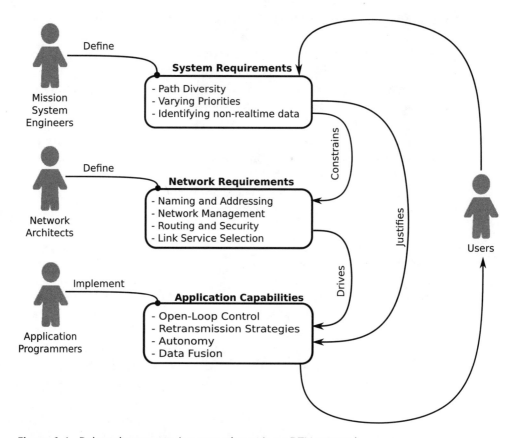

Figure 1.4 Delay tolerance requires more than using a DTN protocol.

requiring constant, high-rate networking access can increase software reusability, reducing cost and improving schedule associated with application development in challenged networking environments.

Finally, users, such as mission operators, use networks and network applications to accomplish mission-specific tasks. This book details examples of the impact of a delay-tolerant CONOP on mission performance, particularly in areas where node management ultimately equates to node autonomy.

1.5 How to Use This Book

This book consists of two parts that address the motivation and design of delay-tolerant networks and design patterns and examples of delay-tolerant application development and deployment.

The first part (Chapters 2–7) discusses the emergence of challenged networking as an expected consequence of wanting to more fully instrument our world and our solar system and the ways in which networking technologies must adapt to enable this communication. Cases where the terrestrial internet encounters problems similar to those that define challenged networks are identified and standard solutions to those problems are presented. An overview of the history of delay-tolerant networking as a solution to challenged networking—and the motivating use cases for DTN—are discussed. Finally, the section concludes with the delay-tolerant networking architecture defining modern DTN and its enabling transport protocol, the BP.

The second part (Chapters 8–17) presents advanced networking concepts associated with DTNs and proposes the use of network design patterns as a way to reason about these different concepts for a specific deployment. This work is motivated by previous efforts to describe software design patterns [12] and qualitative measures to assess compliance with best practices (such as code smells) [13–15]. Having introduced the concept of a protocol design pattern, a series of chapters on specific design patterns for delay-tolerant applications follow. These chapters include how to cache content in the network, perform open-loop autonomous control of nodes, annotate messages to reduce traffic needs, perform distributed error correction, in-network data fusion, and regional administration. The part ends with a discussion on special considerations unique to DTNs that must be accommodated by delay-tolerant applications, examples of using these patterns, and case studies for their deployment.

1.6 Summary

We live in an increasingly connected, wireless world whose infrastructure provides rapid, reliable communications among people and machines. Much of our daily data consumption has come to assume this type of ubiquitous networking and a significant portion of our social and civic communications rely on these capabilities. As such, the majority of terrestrial networking research, development, and maintenance is focused on making networks faster in the face of increasing data volumes.

This networking ubiquity has also led to a desire to instrument all parts of our world and our solar system, from buildings to oceans to remote locations to deep space. However, modern networking protocols fail to operate in these types of challenged networking conditions. To solve these problems, network engineers from industry, academia, and government developed a new networking capability, DTN, to establish functioning networks in places that cannot run traditional, terrestrial IPs.

The advent of DTN capabilities marks the start, not the finish, of solving the problem of challenged communications. Delay-intolerant applications will fail to function in these environments in the same way that delay-intolerant transport protocols fail. Network architects, engineers, developers, and operators must adapt and adopt to the concept to delay-tolerant applications to run over these new delay-tolerant networks.

1.7 Problems

1.1 Provide three reasons why the efficiency metric of goodput/throughput may give an incomplete view of the performance of a network.

1.2 What does the term expediency mean as it relates to end-to-end networking? Provide three possible interpretations of efficiency in this context.

1.3 In a common home wireless network, identify three types of data that have strict performance requirements and three types of data that do not have strict performance requirements. A strict performance requirement can be interpreted as a requirement for data to be neither early nor late by more than two seconds.

1.4 Is the common networking application "ping" a delay-tolerant application? Why or why not? Provide at least two reasons.

1.5 Provide three requirements from the point of view of a systems engineering team that would levy requirements on a network architecting team. Provide three network requirements from the point of view of a network architecture team. Finally, provide three application tools that must be developed by application developers to implement services on the network in service of a mission.

1.6 Explain why using DTN protocols is insufficient to have applications operate across a DTN. Give an example of an application that would fail to operate in a challenged environment even if the network if a DTN.

References

[1] Cerf, V., et. al., "Delay Tolerant Networking Architecture," *RFC4838*, April 2007, DOI 10.17487/RFC4838, <https://www.rfc-editor.org/info/rfc4838>.

[2] Soumya, S., et al. "A Survey of Smart Data Pricing: Past Proposals, Current Plans, and Future Trends," *ACM Computing Surveys (csur)*, Vol. 46, No. 2, Article 15, 2013.

[3] Birrane, E. J., D. J. Copeland, and M. G. Ryschkewitsch, "The Path to Space-Terrestrial Internetworking," *2017 IEEE International Conference on Wireless for Space and Extreme Environments (WiSEE)*, Montreal, Quebec, 2017, pp. 134–139.

[4] Fall K., and S. Farrell, "DTN: An Architectural Retrospective," *IEEE Journal on Selected Areas in Communications*, Vol. 26, No. 5, June 2008, pp. 828–836.

[5] Farrell, S., et. al., "When TCP Breaks: Delay- and Disruption- Tolerant Networking," *IEEE Internet Computing*, Vol. 10, No. 4, July-August 2006, pp. 72–78.

[6] Burleigh, S., K. Fall, and E. Birrane, "Bundle Protocol Version 7," Work in Progress, draft-ietf-dtn-bpbis-12.txt, November 2018.

[7] Ott, J., and M. J. Pitkanen, "DTN-Based Content Storage and Retrieval," *2007 IEEE International Symposium on a World of Wireless, Mobile and Multimedia Networks (WOWMOM)*, Espoo, Finland, 2007, pp. 1–7.

[8] Seligman, M., K. Fall, and P. Mundur, "Storage Routing for DTN Congestion Control," *Wirel. Commun. Mob. Comput.*, Vol. 7, 2007, pp.1183–1196.

[9] Ramadas, M., S. Burleigh, and S. Farrell, "Licklider Transmission Protocol—Specification," *RFC5326*, September 2008.

[10] Cello, M., G. Gnecco, M. Marchese, and M. Sanguineti, "Evaluation of the Average Packet Delivery Delay in Highly-Disrupted Networks: The DTN and IP-like Protocol Cases," *IEEE Communications Letters*, Vol. 18, No. 3, March 2014, pp. 519–522.

[11] Caini, C., P. Cornice, R. Firrincieli, and D. Lacamera, "A DTN Approach to Satellite Communications," *IEEE Journal on Selected Areas in Communications*, Vol. 26, No. 5, June 2008, pp. 820–827.

[12] Caini, C., et. al., "TCP, PEP and DTN Performance on Disruptive Satellite Channels," *2009 International Workshop on Satellite and Space Communications*, Tuscany, Italy, 2009, pp. 371–375.

[13] Fowler, M., "Patterns [Software Patterns]," *IEEE Software*, Vol. 20, No. 2, March-April 2003, pp. 56–57.

[14] Khomh, F., M. Di Penta, and Y. Guéhéneuc, "An Exploratory Study of the Impact of Code Smells on Software Change-Proneness," *2009 16th Working Conference on Reverse Engineering (WCRE)*, Lille, France, 2009, pp. 75–84.

[15] Anda, B. C., et. al., "Quantifying the Effect of Code Smells on Maintenance Effort," *IEEE Transactions on Software Engineering*, Vol. 39, pp. 1144–1156, 2013.

CHAPTER 2
A Brief History of Challenged Networking Environments

This chapter describes the concept of a challenged networking environment and explains how such environments emerge as the natural consequence of communicating beyond the limits of terrestrial infrastructure. The set of networking protocols and applications colloquially referred to as DTN is proposed as the architecture that enables useful communications for these environments. Since the motivation for developing delay-tolerant applications is based on the constraints of operating in a DTN, a discussion of why we have challenged networking environments (and thus need DTNs) is a natural starting point.

2.1 What is a Challenged Networking Environment?

A challenged networking environment is any that impedes reliable and timely data exchange. While this characterization can seem precise, it is predicated upon the terms reliable and timely, each of which can have multiple definitions. The term reliability can be defined in an absolute way such as never having an interruption longer than a certain duration. Alternatively, reliability can be defined relative to some other measurement, like overall uptime (such as the five nines metric [1]) or minimum amount of user data that must be delivered. Similarly, timeliness can be defined in an absolute sense as no message taking longer than a certain number of seconds to be delivered, or it can be defined in a relative sense such as every message being delivered prior to its expiration date.

Absolute definitions for reliability and timeliness are difficult to justify in a networking context. As an example, a web server that takes 2 seconds to retrieve a complex web page might be considered speedy whereas that same 2 seconds applied to a car engine control network may be orders of magnitude too slow. Useful analysis of a network must occur in the context of the purpose and function of the network, what the network needs to accomplish, and its unique performance requirements. As such, absolute interpretations for reliability and timeliness are indefensible. To make a judgment as to whether a network is being challenged by its environment, the environment must be examined in the context of the responsibilities and performance requirements of the network itself.

2.1.1 Separating Responsibilities

Just as terms like reliable and timely are subjective, the term network is overloaded with multiple equally valid definitions. The components of a network may be different depending on whether it is being defined by hardware, firmware, software, or some combination of these. Architects, algorithm designers, and application developers may also assume the presence of different networking capabilities. Therefore, any meaningful discussion of networking assumptions and characteristics must start with an unambiguous definition of what does and does not constitute a network.

One way to accomplish this is to define a network as a series of disambiguated services each with their own behaviors and functional responsibilities. As an example, the network service of modulating physical media for the transmission of information bits is easily distinguished from the network service of determining whether a series of bits must be retransmitted. These are two very different functions. The relationships between and among network services can be integrated through a number of architectural views, with two popular architectural views being layering and encapsulation.

As views of the same information, layers and encapsulations can both be used to discuss network services. Their relationship, and typical graphical representation, is illustrated in Figure 2.1. In this figure, lower layers represent syntax-related services such as encoding and transmission/receipt of information elements through a physical medium. Higher layers represent semantic-related services such as how user information should affect the actions of an application. Lower layers exchange bits whereas upper layers infer meaning from those bits. Since bits encoded as modulation symbols are the actual information elements communicated through

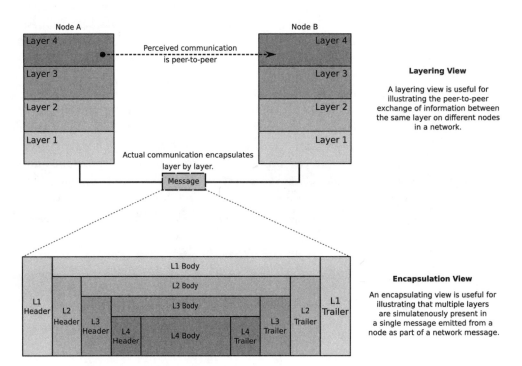

Figure 2.1 Different architectural views of layering and encapsulation.

the physical medium, the information necessary for their encoding and decoding must be the first set of information encountered by transmitters and receivers in the network. Therefore, lower layers form the outer wrappings of encapsulated information, and higher layers form the inner wrapped information being communicated. These two views both provide meaningful representations of network services while simultaneously breaking the interfaces and interactions between those services into manageable pieces.

2.1.2 Defining Network Service Layers

Both views assume the ordering of layers/encapsulations as well defined, deterministic and mandatory; meaning that messages do not skip over or reorder layers or encapsulations. However, no single set of network services can accommodate every network deployment. There exist multiple networking models that each define their own collections of network services and number of layers (typically in the range of three to eight)[1] [2–4]. Network models with several layers tend to have a high-fidelity decomposition of applications/protocols to networking services. Models with fewer layers support more flexibility and allow different combinations of protocols.

We consider a simplified view of networking consisting of three layers: link, network, and application, with each layer supporting multiple protocols as described in Table 2.1.

Layering and encapsulations provide a more precise way to consider the ways in which the network can be challenged by its environment. We can generally state that a challenged networking environment is one that impedes the successful

Table 2.1 A Three-Layer Model Captures the Minimal Set of Network Services

Layer	Typical Layer Services
Link	Detecting presence and loss of links.
	Encode and decode individual bits of information.
	Applying transmission-related error corrections.
	Physically transmit information and/or accumulate information at analog-to-digital boundaries.
	Secure communications links.
Network	Reassembly of information fragmented by the network.
	In-order delivery of information.
	Requesting retransmissions of individual pieces of data
	Determine route information for user data.
	Manage network resources for delivery.
	Secure data communications.
Application	Construct end-to-end user information.
	Determine the recipient for critical information.
	Provide processing information and security specifically for user data.
	Secure information exchange.

---

1. Multiple, nonstandard interpretations exist for supplementary, nontechnical networking layers capturing concerns such as political, user, financial, and regulatory information. Nontechnical layers are not considered in this analysis.

operation of one or more core network services. Following that idea, we can state that a network service can be considered impeded if it is unable to perform any of its functional responsibilities in a timely or reliable way (for whatever definitions of timely and reliable are being used in this particular network context). The general classes of impediments experienced by the three layers defined in Table 2.1 are discussed in the next sections.

2.2 Link Layer Challenges

There are three motivating features associated with the establishment of communications links between two nodes in a network: wireless operation, node mobility, and link heterogeneity. While each of these features allow for data exchange in increasingly dynamic and geographically remote scenarios, they complicate the types of services that must be accomplished at this layer and increase the chances of impeded operation.

2.2.1 High-Rate Wireless Communications

Wireless links have clear advantages over wired links in certain environments as they allow flexibility in the physical placement of networking nodes with less infrastructure investment than is the case with physically wired connections. For this reason, wireless communications are becoming the preferred choice for the last hop of home networks and in some cases the last mile provisioning networking services into the home [5]. A familiar example of last hop wireless is the model used by many internet service providers (ISPs) who provide a wireless router as a termination point for a wired connection to the terrestrial internet. Similarly, in a cellular Long Term Evolution (LTE) architecture [6], mobile phones necessitate the use of wireless links as the last mile for delivering terrestrial internet through an otherwise wired set of connections through a cellular provider's network.

The existence of wireless links is certainly not new. However, their increased use as a replacement for high rate wired links is unique over the past decade. Using cellular systems as an example, the evolution from first to 4th-generation standards was brought about by the need to scale wireless link technology, particularly as it pertained to radio frequency (RF) waveforms, multiple access strategies, and roaming across different service provider networks [7]. Similarly, advances in media access control and physical layers from 802.11(b/c) to 802.11(n) and the selection of 5 Ghz versus 2.4 Ghz spectrum has enabled high rate, reliable data exchange that is competitive with home wired links. In fact, it is often the case that home local wireless links provide higher data rates than the wired links connecting them to the internet at large. Within the past 5 years, a majority of social information consumption occurs over last mile wireless links [8].

While wireless links have certain clear advantages, they also have susceptibilities that are less of an impact to traditionally wired networks. Wireless links have an obviously greater susceptibility to path attenuations than wired links. It is obvious to the network designer or engineer familiar with RF communications that wireless links can be attenuated by atmospheric, RF noise, passing through structure such as walls, and other factors resulting from the physical environment.

Wired links, conversely, pipe through these physical environmental effects and are not significantly degraded.

How wireless networks are deployed can also introduce more subtle weaknesses related to power utilization. The same freedom from physical infrastructure that allows freedom of placement of wireless nodes means that it is less likely that the wireless nodes are placed where there is a regular mains power source. This is problematic given the power required for sustained wireless transmission. While optimizations in wireless adapters allow for very low power idle modes, wireless devices consume significant power (in a relative sense) when they are exchanging continuous data. For example, a wireless outdoor surveillance camera battery may last months per charge with intermittent recording and yet be drained to nothing after only a few hours of continuous recording. In the sense of power constraints, we can say that challenged environments for wireless communications are those that require so much power consumption as to impede the transmission of data.

There is additional interplay between RF link performance and power consumption. When signal attenuation becomes too high (as can be caused by noise in the environment, path loss over long distances, and in some cases, malicious data [9]), a wireless link must expend more power to either constantly retransmit lost data or overcome impairments and maintain data communications. The alternative to expending more power is to accept the failure of the link, causing another form of network disruption. When wireless nodes run on batteries or harvest energy from their environment, they may be unable to transmit regularly as a function of power management and scheduled transmission opportunities may be constrained if too much power is needed to overcome noise in the transmission environment. Real implementations of wireless systems are, of course, physically limited in terms of the power they are able to transmit, meaning that even with increased power consumption there remains a physical limit beyond which the transmitter cannot overcome its environmental interference. Similarly, real-world receivers have limited sensitivities. Continuously increasing transmit power to overcome channel noise will eventually drive the combined signal beyond the receiver's ability to receive either by overdriving the receiver front end (saturation) or by raising the noise floor such that signals cannot be distinguished from noise.

For example, consider a brief analysis of the capabilities of various IEEE 802.11 wireless link standards [10] shown in Table 2.2. This table lists an estimate of the speed of each standard in Mbps, and the nominal indoor range, in meters. This table demonstrates the multiple challenges involved in balancing concerns relating to a wireless link. Very high rate transmissions, such as those achieved in 802.11(ac) and 802.11(ad), require high power and can only be reasonably achieved over short distances. 802.11(ah), which operates over a much longer range of 1 km, achieves this by moving to a much lower frequency and combining multiple channels to achieve faster data rates. In its most interoperable mode, it achieves only 4 Mbps and cannot get close to the speeds achieved by modern wireless standards over shorter distances.

It should be noted that a network designer can alter the environment and change the rules by which a protocol must operate. In the case of a high-speed protocol such as 802.11(ac), which was designed for short distances but assumes a near omnidirectional channel path, the addition of directional antennas can provide channel gain to overcome distance at the expense of limiting the directionality

Table 2.2 Rate and Range Are Functions of Power Availability

802.11 Standards	Year	Frequency (Ghz)	Maximum Speed (Mbps)	Indoor Range (m)
b	1999	2.4	11	~40
g	2003	2.4	54	~40
n	2009	2.4/5	600	70
ac	2013	5	3466	~35
ad	2012	60	~7000	3.3
ah	2017	0.9	347	1000

of the communication. The designer is trading directionality for data rate and link distance. He or she has changed the rules in order to modify the environment in which the protocol will operate and therefore achieve a network function not otherwise available. The takeaway is that a network designer should not necessarily limit themselves to the general assumptions of a given protocol. By changing the environment, they can change the performance of the network.

When discussing wireless link impairments, expected performance is used to calculate the relative measure of the network. For example, achieving a maximum throughput of only 1000 Mbps over 802.11(ad) links would indicate some type of impairment whereas achieving a third of that throughput over 802.11(ah) links would represent near perfect performance. Similarly, attempting to transmit across 10m in 802.11(ad) would result in link loss, whereas 802.11(ah) would transmit across 100 times that distance.

2.2.2 Node Mobility

One reason that wireless link specifications consider range is the recognition that transceivers on either end of the wireless link may be in motion. There are three ways in which environments that support node mobility can impede data transmission, as illustrated in Figure 2.2.

Figure 2.2 illustrates the hub-and-spoke networking architecture commonly found in home networks. This architecture comprises a single, stationary access point and a variety of devices (some mobile) that connect to that access point. Mobility creates no impairment if the moving device is within the range of the access point and without sources of interference. When a device moves far enough away, the link rapidly degrades, resulting in degradation of the data channel. This is a familiar limitation experienced by consumers whose last hop is between a wireless access point and their wireless device (a laptop or a cellphone). Buildings often have dead zones or other areas of limited wireless coverage resulting from physical and electromagnetic blockages and access point placement can be an important consideration in setting up a home wireless network. When a single access point cannot cover the practical area within which a node may move, techniques must be used to coordinate coverage of the mobility area. Devices such as Wi-Fi extenders and mesh routers can be placed throughout a mobility area to eliminate dead zones in larger structures.

This approach of populating access points across a mobility area scales well even when the mobility area is quite large. For example, cellular network providers

2.2 Link Layer Challenges

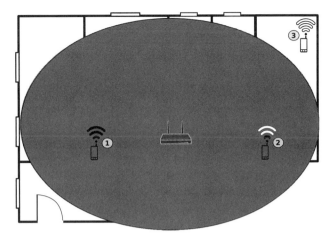

Mobility and Impairments

Even in controlled environments node mobility significantly impacts link performance.

① Devices close to each other with few obstacles can have excellent communications.

② Devices close to each other with some obstacles can have degraded communications.

③ Devices that move far enough away from each other lose all communications.

Figure 2.2 Node mobility disrupts wireless links as a function of range and directionality.

distribute tens of thousands of cellular access points (towers) across thousands of miles to ensure that highly mobile cellular phones have uninterrupted service. While accurate counts of cellular towers deployed in the United States are difficult to attain, the number of such sites in 2018 listed as registered and active by the FCC exceed 120,000 [11] and will continue to grow as a function of new users and new antennas to support evolving link layers for 5G systems and beyond.

To illustrate this, consider a cell phone traveling along a highway, moving out of the range of one cellular tower and into range of another. A process called handoff allows one tower to negotiate transition of the wireless link from a given cell phone to another tower, as shown in Figure 2.3. From the cell phone user's perspective, the area of wireless coverage is seamless. While coordinated handoffs have become a reliable way of keeping uninterrupted cellular service it can become complex to negotiate in areas with many overlapping access points or when nodes are moving so fast that they traverse multiple access points faster than those access points can perform handoff.

Several scenarios exist where it is not practical to provide access point coverage across a large mobility area either because of the cost of placing sufficient access points is prohibitive or because nodes are moving too fast to negotiate handoff. For example, cell phones traveling in cars down a highway can be serviced by traditional cellular towers whereas cell phones traveling in airplanes in flight are challenged because of both the speed of the airplane and its relative distance from the towers.

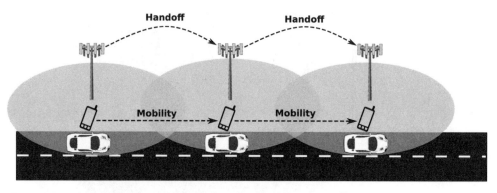

Figure 2.3 Adding access points reduces mobility issues.

Figure 2.4 illustrates the situation where both user nodes and access points are mobile, as is commonly the case in proposed space networks, vehicular networks, and general ad hoc networks. In this example, air traffic is unable to use existing networks of ground-based cellular towers for in-flight communications and relies on low Earth orbit (LEO) spacecraft to perform a similar function. The in-flight Wi-Fi experience is typically different and less interactive than in a home or office as a function of the relative sparse connectivity and propagation delay of available spacecraft links.

In all cases, mobility provides a way to preserve network communications across a larger area without needing to densely populate the area with nodes. Rather than having a desktop computer in every room of a house, laptops and smartphones allow for a similar experience with much less investment. Spacecraft more

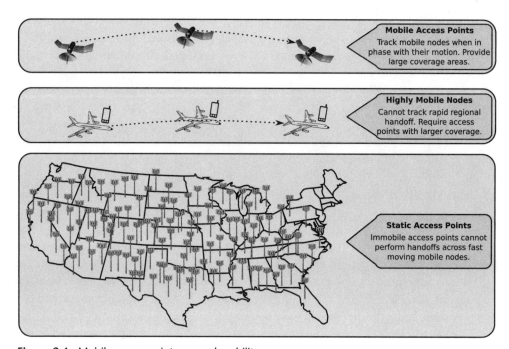

Figure 2.4 Mobile access points expand mobility area coverage.

economically provide coverage to rural and otherwise unpopulated areas, such as deserts and oceans, than either cellular or wired infrastructure.

When mobile nodes are used to provide connectivity in a sparsely populated network, their mobility has the potential to partition the network into disconnected segments as links are established and dropped over time. In cases where the mobility pattern is regular, techniques can be used to schedule data transmissions in advance. When the mobility pattern is irregular, it becomes much more difficult to plan for future link availability and data may be unable to be communicated in a timely or reliable manner (again depending on our definitions of timely and reliable).

2.2.3 Link Heterogeneity

As computing devices become more capable, their ability to contribute as nodes in the network becomes more complex. Early wireless deployments made simplifying assumptions regarding the capabilities of their nodes, to include the assumption that all networking devices support the same physical link layers. More recently, network nodes support a variety of link layers whose configurations are controllable by local conditions or remote administration. This can be accomplished using flexible, software-defined radios that provide multiple functions by changing configurations or by cohosting multiple simple physical radios in a single device. As an example, modern smartphones incorporate separate radios for Wi-Fi, Bluetooth, near-field communication (NFC), and cellular (4G/5G/LTE).

Different types of data links in a network provide flexibility in terms of coverage areas, data rates, and cost. A user may restrict smartphone applications to Wi-Fi service to avoid cellular service fees, and later, restrict smartphone applications to cellular service to avoid using an unsecured public Wi-Fi network. In both cases, this user benefits from the flexibility of multiple radios and link diversity (as a function of coverage, speed, and pricing). Beyond providing different connectivity choices on a node-by-node basis, nodes that support heterogeneous links can serve as gateway or border nodes federating individual networks into a single internetwork. For example, a smartphone operating as a hot spot can serve as a gateway between a local Wi-Fi network and a cellular network.

The flexibility provided by link heterogeneity comes with an increase in configuration cost and complexity. There are situations where nodes in a mixed network may be in close physical proximity but unable to communicate because of an incompatible link layer configuration. Nodes having multiple radios may not be able to utilize them concurrently due to issues with interference, pointing, or power constraints. When multiple radios can be used simultaneously, nodes must make the determination of which radios are best utilized for which traffic. Configuration issues can prevent link establishment, create bottlenecks, and require buffering and queueing strategies in much the same way as links lost to node mobility. Generally, the more complex the configuration the more likely it is that the network can be misconfigured or configured in a way that reduces overall throughput or decreases security. The end result is that poorly configured link heterogeneity can cause data to not be delivered in a timely or reliable manner.

2.3 Network Layer Challenges

A network is not a random collection of links between nodes; it is a coordinated collection of links between nodes, where coordination takes the form of asserted state information, performance measurements, and synchronization. The services that form a network by coordinating available links are grouped in the network layer. There are three typical ways in which an environment can impede the function of services in this layer: path losses, time-variant topologies, and unsynchronized nodes.

2.3.1 Path Losses

Data, as a sequence of packets, pass through a network via a sequence of links. Any packet may be lost during transmission across any individual link as a result of the kinds of link layer challenges discussed in the prior section. We define the resulting path loss as the sum of all packet losses of all links comprising some end-to-end path. As an assistive measure, path loss can be significantly higher than any individual link loss. Correcting for path loss can also be much harder based on the types of links comprising the path, the speed and volume of data, the number of packets, how many other paths might be serviced by common links, quality of service considerations, and a variety of other metrics. Networks that assume highly reliable, highly available links may not have an effective way to measure the magnitude of, or recover from, significant path loss.

To understand how path loss can be measured, one needs to understand how individual link loss is measured. Two metrics used to measure link loss are the bit error rate (BER) and packet error rate (PER). BER is the probability that any given bit received over the link is in error, whereas PER is the probability that any given packet received over the link contains an error. For example, if a link has a data rate of 100 kbps and a BER of 10^{-5}, then 1 bit would be received in error every second. The relationship between BER and PER contains subtle implications based on packet size and the fact that transmission protocols retransmit entire packets and not portions of packets. In the prior example, while there may only be one erroneous bit over a link per second, if that one error appears in a 1,500-byte packet, then the entire 12,000 bits of that packet must be retransmitted, not just the one bit that was in error. The solution of retransmitting any packet that experiences even a single bit of error is unacceptably wasteful. Since the start of wireless transmission, strategies to over-sample data to try and fix small errors have been deployed to make networks usable [12, 13]. These error correcting codes attempt to reduce BER (and, thus, PER) by redundantly encoding portions of data such that flipped bits can be detected and corrected by a receiver. This redundant encoding can include parity bits, checksums, or multiple copies of the data. The selection of correcting code is often based on the nature of the data being corrected and the likely causes of errors over a link. Advances in these coding techniques have reached close to the theoretical limits of their performance [14] for RF links and new work continues for their application to higher rate links, such as those associated with optical communications [15]. While error correcting codes fundamentally enable communications—and particularly wireless communications—BERs are never zero.

While the probability of a single link experiencing data loss may be low, the probability that an end-to-end path experiences data loss grows with the number of links comprising the path. Even with error correcting codes, PERs cannot be completely avoided and packet retransmission will always be a necessary capability. The best way to reduce the impact of packet retransmission is to reduce the size of individual packets; smaller packets require fewer network resources to retransmit (in terms of any given single packet). However, small packet sizes mean that the relative size of the packet overhead is greater, and therefore the overall efficiency of each packet is lower (fixed-sized packet headers and metadata are a larger percentage of a smaller packet) and keeping a low BER implies power, proximity, and other resources be available at each node in the network.

Combating path losses becomes increasingly difficult as paths are longer, as links experience more disruptions, as heterogenous links experience different kinds of impairments, and as smaller packet sizes reduce overall network efficiency. Failure to account for these issues will prevent the network layer from providing timely and reliable data exchanges.

2.3.2 Time-Variant, Partitioning Topologies

The instantaneous topology of a network changes as individual links in the network come and go as a function of the types of link and network challenges discussed so far in this chapter. This means that as the topology of the network varies over time, the network may exist in multiple partitions as a result of this topological change. While this level of change is apparent in any network to some degree, challenged networking environments are unique in that time-variant change and network partitions can become the normal situation rather than a transient error condition.

In some cases, mobility causes link impairments by increasing the distance between nodes. In other cases, mobility preserves links by maintaining a minimum distance between nodes or moving to heal portions of a network when other nodes experience failures. While node mobility can preserve individual links, time-variant topologies make it difficult or impossible to coordinate individual links to make predictable paths through the network. If node mobility partitions the network or changes the links that comprise an end-to-end path, the network layer may be unable to deliver end-to-end user traffic in a timely and reliable manner without adapting to the changes. This is clearly the case if mobility is random but can also be the result of planned topological changes if those changes prevent timely data transmission.

2.3.3 Unsynchronized Node Information

Because a network is a coordinated series of links, one responsibility of the network layer is performing that coordination. In this context, coordination means configuring nodes in the network to perform the same (or compatible) activities at the same time or in some required sequence. This type of coordination requires individual nodes to have some common sense of time and state that can be difficult or impossible to achieve in a challenged networking environment for three reasons.

First, nodes without regular access to a mains power supply may periodically shut down to conserve energy or recharge batteries. When they reboot, having an accurate sense of time and network state becomes difficult without synchronizing with a device that was maintaining this information while they were dormant. Second, as a function of cost, some nodes (such as those used in large distributed sensor networks) may have imprecise oscillators resulting in clock drift over time. Third, nodes in a network may need to correct for signal propagation delays based on their performance requirements. For example, nodes separated by interplanetary distances may incur minutes to hours of propagation delay making timing synchronization difficult as a transmitter may need to transmit a signal to where a receiver can accept it in the future, essentially leading the target as the bits cross interplanetary space. Such links are often unidirectional, and thus, unable to synchronize information effectively.

For these reasons, applications that require tightly coordinated time at each node will fail. Typically, applications with these types of requirements are not user applications, but those that implement the protocols and behaviors necessary for the network itself to function. If the protocols and behaviors upon which the network relies encounter failures, the network itself will fail.

2.4 Application Layer Challenges

The characteristics of challenged networking environments, whether at the link layer or network layer, have the effect of impeding data exchange between programs operating at the application layer. There are three types of application-layer issues that stem from these lower-layer impediments: growing delays, increased disruptions, and increased bursts of data volume.

2.4.1 Growing Delays

A delay at the application layer is the amount of time that an application must wait between requesting some action from the network (a data query, a status request, or the running of some remote command) and the time at which the action occurs (data or acknowledgements are received, a command is run). All network applications experience some delay, where delays are caused by either signal propagation and/or by retransmission caused by data loss.

Common application delays on the terrestrial internet are measured in milliseconds with delays of one or two seconds being considered the limit of acceptable performance. As network applications operate in more challenged networks, the magnitude of delays will increase. Increasingly distributed deployments (to include astronomical distances) increase signal propagation delays. More frequent link disruptions caused by temporary loss of signal (such as nodes that move too far from each other) or randomly transient (such as when operating in a noisy or contested environment) require detection and data retransmission, which will also result in longer application delays.

Delays become problematic when applications need to make time-sensitive decisions such as querying the network for information as part of performing a common networking operation. For example, it is a common occurrence for a node on the terrestrial internet to check the validity of a security certificate with a remote certificate authority as part of establishing a trusted communications session. That same application operating in a challenged network may not have timely access to such an authority and would be unable to perform the sequence of actions necessary to establish a communications session.

2.4.2 Increased Disruptions

A disruption at the application layer is one that occurs over any link between a source and destination application communicating over the network. These disruptions can occur at the node where the source application is running, at the node where the destination application is running, or on any node passing source application data. When networked applications developed for the terrestrial internet run in a challenged network environment, they will experience an increase in the frequency and magnitude of disruptions.

When individual link disruptions are infrequent and short-lived, retransmission strategies are effective, and disruptions only increase the overall application delay in the system. However, when these disruptions are frequent and/or persist for long periods of time, the incurred delays may exceed application timeouts and appear as application disruptions.

Applications operating in challenged environments are more likely to experience disruptions because the individual links forming the application communications path are themselves more likely to experience disruptions. This often requires a departure from traditional architectures that rely on the concept of a dedicated end-to-end connection between applications, such as the familiar model of a TCP/IP session over the terrestrial internet. If a node cannot establish a session because the challenged environment is constantly partitioning the network, then a traditional application may never pass any data at all. In addition to a reduced likelihood of success, healing approaches based on repeatedly trying to reestablish sessions can never truly mitigate the underlying disruption and only wastes networking and power resources, which can be problematic on resource-constrained nodes.

2.4.3 Growing Data Volumes

Applications on the terrestrial internet may use a variety of network and application specific mechanisms to determine how fast to inject data into the network. Network flow control mechanisms, service-level agreements, quotas, back-pressure protocols, and other techniques allow high-rate, high-availability networks to tweak the nature of data exchange. In challenged networking environments, this control over the application data volume is difficult for two reasons: timely data exchange and in-network storage.

In the presence of delays and disruptions, applications have no guarantee that any measurement of data flow (or request for data throttling) will be received by a transmitting application in time to prevent communications problems. As such, the overall data volume from a transmitting application may grow to be very large absent of any way to request that the application send data slower. Even in the case where throttling messages are communicated, they may not be received in time to have the necessary effect. Such closed-loop network traffic control protocols are fundamentally susceptible to increased delay and disruption in the underlying links.

Even if a transmitting application produces data at a reasonable rate, there is no guarantee that a network operating in a challenged environment will be able to deliver that data without delay. When information is delayed in the network, it must be buffered somewhere while waiting for an appropriate transmission opportunity. That somewhere is a real network node, with real storage systems, having real constraints. It is all too likely that too much data will accumulate at a node forcing the network to either proactively discard information or suffer the consequences of overloading. When a transmission opportunity presents itself to a storage node, there will be a burst of activity as the buffers are cleared. Similarly, when a network node is partitioned from the network, any data that has been queued for transmission will be communicated as quickly as possible. In cases where network connectivity is intermittent, large bursts of data volumes can saturate links, choking out new information that is attempting to transit the network.

2.5 Error Handling in Challenged Networking Environments

No network is perfect, and any resourced (high availability, high reliability) institutional, cellular, or home network experiences occasional impairments (delays and disruptions). What differentiates resourced networking environments from challenged networking environments is the frequency and magnitude of these impairments and how they must therefore be handled. In resourced networking environments, impairments are treated as error conditions for which there are recovery procedures. In challenged networking environments, impairments are normal operating behavior and there is no additional recovery process separate from the message passing process. The network is designed assuming there will be impairments as the normal course of business, therefore there is nothing to recover from.

The determination of what is normal operating behavior versus what is an error condition is a fundamental consideration when designing a networking application. Networking applications assume that networking protocols handle network error conditions. When an impairment is treated as an error condition in a resourced environment and not treated as an error condition in a challenged networking environment, then the same network application design cannot be used in both environments.

2.6 What Is a Network Error Condition?

The most succinct definition of a network error condition is a condition that requires recovery. If a network is experiencing an error, it is not operating as it should and some type of action must be taken to get the network back to its proper operation. Resourced networks treat impairments as errors because they are assumed to be short-lived, infrequent, and/or predictable. Short-lived and infrequent impairments can be recovered by waiting for the network to heal and then trying again. Predictable impairments can be recovered by waiting until a specific time and then trying again. In either case, resourced networks heal by detecting the error, waiting for it to pass, and then reestablishing connectivity.

This type of recovery is inappropriate when impairments are frequent or long-lived. For example, consider a common resourced network function: creating and using a TCP/IP session between two nodes—Node A and Node B. This involves a session establishment phase, a data exchange phase, and a session cleanup phase. When communications are lost, the session is lost, and the underlying networking protocol refuses to accept new user data while it recovers by attempting to reestablish the session. Once the session has been reestablished, then user data can be exchanged.

This approach to recovery is sensitive to the frequency and duration of the network impairments. If impairments are short-lived enough, they fit within the regular timing of the network, the applications detect no errors at all, and no recovery is required. Long but infrequent disruptions may cause loss of data or loss of session, but because these errors are infrequent the network can recover and then proceed as normal. However, when impairments are frequent (and possibly long), attempting to enter a recovery state may never result in data exchange; if the recovery process takes longer than the period of network connectivity, then all opportunities to exchange any kind of data are wasted by attempting to recovery the session rather than passing user data. This situation is shown in Figure 2.5.

Because challenged networking environments exhibit frequent and/or long-lived impairments, this type of recovery process cannot be used. For similar reasons, networks in these environments cannot support any type of special recovery based on the presence of impairments; if a network required recovery every time there were significant delays and disruptions then it would be in a perpetual state of recovery and never pass data at all. A different way of handling impairments is needed.

2.7 Approaches to Handling Error Conditions

Absent a dedicated recovery procedure, the most popular approach to dealing with impairments is to only build networks in well-resourced environments. Alternatively, great expense is often incurred to try and transform challenged environments into resourced environments. This transformation is done be either adding more nodes, speeding up existing links in the system, and/or increasing the efficiency of protocols used to construct the network. The underlying argument is that if you throw enough resources at a problem, the problem will cease to be a problem. We will see that this is not always the case.

2.7.1 Increase the Number of Nodes

Many impairments associated with challenged environments stems from sparse population of the network. Adding more nodes can increase the average number of links in the network if the added nodes are properly placed. This usually means placing nodes in areas of existing network connectivity to increase the robustness of the network rather than placing nodes in areas with no existing connectivity to extend the range of the network. Increased connectivity then implies fewer impairments in the network.

The concept of connectivity comes from graph theory. When networks are modeled as a graph, connectivity refers to the minimum number of nodes that can be removed from a network before end-to-end communication between any pair of remaining nodes is made impossible. For example, in a fully connected network, every node is connected to every other node. Removing a single node does not endanger the ability for the remaining nodes to communicate. In such a case, the connectivity of the network is N-1. This means that all nodes except for 1 need to be removed from the network before communication between any remaining pair of nodes is made impossible because, after removing N-1 nodes, only one node remains in the network! Often, networks try to maintain 2-, 3-, or 4-connectivity, meaning that there exists some redundancy in the network such that it can survive the loss of some number of nodes.

Maintaining high connectivity in densely populated regions with ready access to supporting infrastructure is rarely an issue. For example, modern cities support hundreds of cellular towers representing high coverage area (and high EMF interference) because the cost of cellular towers is relatively small whereas areas with less infrastructure require more significant investment. The ocean cannot be practically populated with networking ships and buoys. The desert cannot be practically populated with cellular towers. Near-Earth, cislunar, and interplanetary space cannot practically be densely populated with spacecraft.[2]

2.7.2 Alter Link Characteristics

Changing link characteristics and/or data rates enable more regular data exchange in networks. Using different frequencies, such as in channelized links, or algorithmically reusing spectrum as in spread-spectrum techniques, increases the robustness of node links to environmental impairments. Both frequency diversity and algorithmic channel reuse improve link robustness in the presence of noise (environmental or intentional) and allow greater bandwidth to occupy the same spectrum real estate. Increasing data rates as a function of using different spectrum or more enhanced symbol encoding does not provide significant robustness against sustained environmental impairments such as active jamming or ongoing interference, but does allow data to be communicated in shorter timeframes, which requires links to be available for shorter periods of time allowing some measure of mitigation of transient impairments such as radio noise from short-duration manmade and natural events.

The theoretical maximum throughput of a link is a function of the data rate of the link and the duration of the link (with the theoretical maximum throughput of an always available link being infinity). Higher throughputs benefit networks when there is a burst of data, which can happen in any network but happens frequently in challenged environments as a result of releasing buffered data as new contacts

---

2. As of this writing, there are several companies seeking to build LEO megaconstellations of spacecraft in near-Earth orbit, based on the (yet unrealized) promise of billions of dollars of profits in an emerging, globally ubiquitous internet marketplace. Even these megaconstellations are faced with resource scarcity in the form of orbital positions, spectrum constraints, and resource-challenged users. It remains to be seen whether these architectures prove to be both technically and economically practical.

2.7 Approaches to Handling Error Conditions

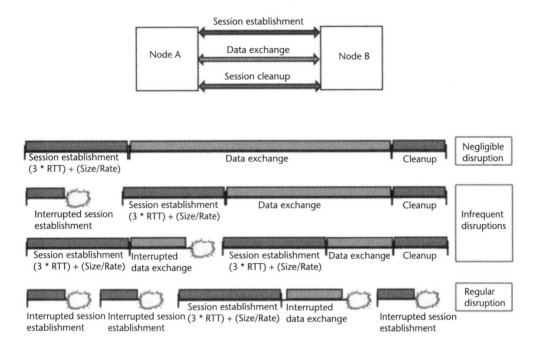

Figure 2.5 Resourced approaches to error recovery cannot scale in challenged networks.

are established over time. Faster, more resilient links will handle this case far more successfully than slower links that may not be able to carry the entire backlog of information. This is the concept of a fat versus a narrow network pipe.

Unlike increasing the number of nodes, link characteristics are unable, by themselves, to solve the problem of network disruptions. Improving the speed of a link will not help in cases where the link is severed. Additionally, there are cases where increased link speed can make network impairments worse, as is the case whereby increasing the rate at which one link in a network path can send data, other links along the path may become congested.

2.7.3 Increase Protocol Efficiency

In addition to creating more links and paths in a network, protocol efficiency allows for more effective use of existing bandwidth. Reducing the size of messages by migrating to more efficient protocol encodings, or by requiring less information in those protocol messages, has the benefit of not only increasing the amount of data that can be pushed through a network, but doing so without requiring structural changes to the network that impact cost and schedule and increase complexity. When discussing protocol efficiency, it is important to differentiate between control plane protocols and data plane protocols.

The control plane of a network consists of the set of messages (and in some instances, physical links) that are dedicated to organizing the information necessary for data exchange. This includes scheduling information, link state monitoring, congestion prediction, routing, management, network-layer security, and other health and status concerns. Efficiencies in the control plane are the most insulated from network application developers but are also the most difficult to optimize as they perform critical functions necessary to maintain data communications.

The data plane of a network consists of the set of messages that carry user application data through the network. The data plane relies on the coordination and information provided by the control plane, but otherwise abstracts the function of the control plane. For example, applications may provide a destination address for their message without concern for the routing protocols, secure session establishment, and maintenance that goes into building and maintaining a path to that destination.

2.8 Summary

Traditional network development focuses on optimizing existing high-availability, high-rate installations such as cellular, home wireless, and the wired terrestrial internet. That focus is not enough to address the explosive growth of networks possessing many characteristics that break the end-to-end connectivity model of the terrestrial internet. With the rise of machine-to-machine interfaces, geographically distributed sensing platforms, and sparsely populated, widely distributed communications needs, a new type of network is emerging that can no longer rely on high-availability, high-rate, omnipresent infrastructure. These networks are termed challenged networks because they differ from traditional networks due to a series of link, network, and application-layer challenges that are either not present, or only present during error conditions, in traditional networks.

Traditional approaches to communicating in challenged networking environments involves transforming the environment itself rather than developing new networking technologies. These approaches include adding resources in the form of additional nodes and/or links into the network to try and provide constant coverage over large geographical areas and the introduction of expensive and exquisite link technologies to greatly increase data rates.

An alternative approach to dealing with such environments is to rethink the real-time nature of data and develop new networking technologies. New protocols and services can be devised to handle the delays and disruptions associated with challenged environments. This problem cannot be completely solved at the network layer. New networks that tolerate delays and disruptions will require new types of applications to run within them.

2.9 Problems

2.1 Provide an example of a challenged network that is operational today. Explain why this network meets the definition of a challenged network.

2.2 Consider a common networking stack of 500 bytes of user data being communicated over a TCP/IP session running in Ethernet frames. Diagram an example encapsulation of the data using typical TCP, IP, and Ethernet headers. In this example calculate the overhead of the overall Ethernet frame relative to the payload size.

2.9 Problems

2.3 Provide an example of a networking application that does not follow the strict separation of responsibilities for network layers identified in Table 2.1.

2.4 Provide three examples of node mobility in common networking applications today.

2.5 Provide an example of using a mobile access point to track a mobile node. Explain why this approach is cost effective.

2.6 Provide three possible reasons why modern networks use links with different link characteristics

2.7 Assume that each link in a network has a 0.5% chance of corrupting the packet such that the packet must be retransmitted. How many hops must the packet take in the network before the chance of packet retransmission is at least 10%?

2.8 Assume that a network path consists of 10 hops, with each hop having a 0.1% probability of causing the packet to be retransmitted from the source. How many times would the packet need to be retransmitted to have at least a 99% chance of delivery?

2.9 Provide two examples where adding nodes to a network is an impractical way to deal with link challenges in the network.

References

[1] Arno, R., P. G. Pe, and R. S. Pe, "What Five 9's Really Means and Managing Expectations," Conference Record of the 2006 IEEE Industry Applications Conference Forty-First IAS Annual Meeting, Tampa, FL, 2006, pp. 270–275.

[2] Zimmermann, H., "OSI Reference Model–The ISO Model of Architecture for Open Systems Interconnection," *IEEE Transactions on Communications*, Vol. 28, No. 4, April 1980, pp. 425–432.

[3] Whitt, R. S., "A Horizontal Leap Forward: Formulating a New Communications Public Policy Framework Based on the Network Layers Model," *Federal Communications Law Journal*, Vol. 56: No. 3, Article 5, 2004.

[4] Feldmann, A., "Internet Clean-Slate Design: What and Why?" *SIGCOMM Comput. Commun. Rev. 37*, No. 3, July 2007, pp. 59–64.

[5] Cherry, S. M., "The Wireless Last Mile," *IEEE Spectrum*, Vol. 40, No. 9, September 2003, pp. 18–22.

[6] Dahlman, E., S., Parkvall, and J. Skold, *4G, LTE-Advanced Pro and the Road to 5G*, London: Academic Press, 2016.

[7] Agrawal, J., et. al., "Evolution of Mobile Communication Network: from 1G to 4G," *International Journal of Multidisciplinary and Current Research*, Vol. 3, November/December, 2015.

[8] Lenhart, A., K. Purcell, A. Smith, and K. Zickuhr, "Social Media & Mobile Internet Use among Teens and Young Adults, Millennials," *Pew Internet & American Life Project*, 2010.

[9] Vasserman, E. Y., and N. Hopper, "Vampire Attacks: Draining Life from Wireless Ad Hoc Sensor Networks," *IEEE Transactions on Mobile Computing*, Vol. 12, No. 2, February 2013, pp. 318–332.

[10] IEEE 802.11 Wireless Local Area Networks Working Group, http://www.ieee802.org/11.

[11] United Stated Federal Communications Commission, "Antenna Structure Registration Search," wireless2.fcc.gov/UlsApp/AsrSearch/asrRegistrationSearch.jsp.

[12] Hamming, R. W., "Error Detecting and Error Correcting Codes," *Bell System Technical Journal*, Vol. 29.2, 1950, pp. 147–160.

[13] MacWilliams, F. J., and N. J. A. Sloane, *The Theory of Error-Correcting Codes*, Elsevier, 1977.

[14] Berrou, C., A. Glavieux, and P. Thitimajshima, "Near Shannon Limit Error-Correcting Coding and Decoding: Turbo-Codes. 1," *Proceedings of ICC '93–IEEE International Conference on Communications*, Geneva, Switzerland, 1993, Vol. 2, pp. 1064–1070.

[15] Tzimpragos, G., et al., "A Survey on FEC Codes for 100 G and Beyond Optical Networks," *IEEE Communications Surveys & Tutorials*, Vol. 18.1, 2016, pp. 209–221.

CHAPTER 3

How the Internet Does It: Approaches and Patterns for Challenged Networking Environments

Several application architectures and protocols must exchange data over the terrestrial internet in situations that are conceptually like those encountered by challenged networking environments. Whereas these environments experience issues due to physical limitations, on the terrestrial internet challenges emerge as a function of scalability and data priority. This chapter reviews approaches to issues found on high-availability, high-speed networks. The unique aspects of these approaches are discussed both in terms of why they work and whether they provide insight into how to solve similar problems in challenged networking environments.

3.1 Challenges in the Terrestrial Internet

While challenged networking environments are most often defined by link discontinuity, limited bandwidth, and node availability, these are not the only impediments to timely data exchange. Even in an otherwise highly available, high-rate network (see Table 3.1), nonphysical phenomena can cause data to be delayed. Network policies, application workflows, pricing, and application behaviors can all result in data delivery delays. This section reviews potential sources of data delays.

3.1.1 Network Policies

Networks employ predetermined management policies in order to provide differentiated services by not treating all traffic in the same way. Traffic shaping mechanisms manipulate how data goes through the network to enforce service agreements [1], prioritize certain types of data transfers [2], and control end-to-end delay [3]. For example, lower priority data in the network may be delayed to permit higher priority data access to a link, with priority determined by how much the sender (or receiver) of the data pays for the network or how critical the data is for network control versus user services. Data may be further differentiated by type with all other factors remaining equal. For example, voice and video streaming is dependent on timely and in-order delivery, while data between applications such as e-mail and file transfer do not have the same requirements. In certain cases, bursts of

Table 3.1 Nonchallenged Environments Also Have Obstacles to Data Delivery

Challenge	Description	Similarity to Challenged Environments
Network Policies	Lower priority data may be delayed or dropped to make way for higher priority data.	Transmitted data may be lost in a noisy environment.
Application Workflows	Data may be cached waiting for pull mechanisms and other query semantics.	Data may be cached at nodes waiting for a new link opportunity.
Costing	Links may be avoided or throttled if they are too costly.	Links may not be present in the network. Data links may be limited or unidirectional due to power, distance, and other bandwidth limiting and availability factors.
Application Behaviors	Application processing may require storing data prior to transmitting.	Data may be cached at nodes waiting for a new link opportunity.

high-priority data prevent lower priority data from being passed in the network prior to its expiration. This is particularly problematic in high-availability networks that assume data packets experience only millisecond delays or in networks providing critical services such as medical monitoring and industrial control systems.

The processes by which network service providers determine and enforce priorities have tremendous impact on the quality of user experiences and the costs associated with use of the network. While a global discussion, these networking policies are most publicly discussed in the United States in the context of network neutrality (or net neutrality)—the concept that data in a network not be treated differently as a function of the source of the data, or its intended purpose. Indeed, the debate has risen to the level of public debate and competing arguments before courts, legislatures, and government regulatory agencies [4, 5]. In cases where data is differentiated based on user, purpose, source address, and other mechanisms, then packets may be delayed or lost as a function of business priority versus link availability. Networks with this level of control can be used to create high-availability networks for some types of data (or some preferred or higher value users) while others experience a much more challenged, DTN-like network.

3.1.2 Data Workflows

Data workflows describe the ways in which applications produce and consume their data in the network. One familiar workflow is to have source applications produce their data and then immediately send that data to its corresponding destination application. However, based on the circumstances under which the data are produced, and the reasons for processing the data at its destination, alternate workflows have evolved.

When the amount of data produced is large relative to the ability of the producer to communicate it, local autonomy schemes can be implemented to reduce the volume of data requiring transmission. As an example, large amounts of sensor data can be summarized and reported at regular intervals, or alternatively, data can be sent in the network only if it is known that a consuming application would want to receive that set of data [6].

Several smartphone application workflows do not provide real-time delivery of data (game status, e-mail, or other notifications) because constant delivery would drain battery life, [7] and in some cases, graphical or audible alerts would be distracting to users. When receiving applications are numerous and distributed, producers may opt for a workflow in which data are cached at strategic locations in the network (in a sense predeploying data) where they can be queried and delivered to consumers in a way that distributes network traffic more equitably than having a data producer communicate directly to every data consumer [8]. Using this model, many home video surveillance systems generate video within a user's home network but store that video on load-balanced servers in the network for consumption by viewing applications that exist in web browsers or smartphone applications. When data caching is used in a network, mechanisms for determining cache sizing, calculating data lifetimes, and avoiding unnecessary duplication must be developed, and these strategies may be like those used when caching information in a challenged networking environment.

3.1.3 Network Pricing and Economics

The price of using a network resource translates into a cost to the user for the use of that network, such that preferring lower priced networking resources incurs a smaller cost on the network user. This can be considered as the financial cost to the user or other resource costs such as the price of storage or energy required to transmit. When a user does not want to pay some cost associated with the consumption of a networking resource, they may find a lower-priced resource currently available or they may wait for another resource later. By relying on general user preferences to minimize cost, network utilization can be shaped as a function of price [9].

A familiar example of (financial) cost delays in a network involves smartphones, where a user may require that certain applications only use Wi-Fi connections instead of cellular connections – an occurrence so frequent that this ability to whitelist applications for cellular service is built into modern phone operating systems. In this instance, Wi-Fi connections are typically seen as incurring a lower cost to the user (or at least to the cellular network provider), either by providing free service or conserving minutes on a remaining cellular plan (or avoiding pay-as-you-go cellular plans) [10]. In these cases, data may be delayed as a function of waiting for a less expensive data option, such as not sending smartphone notifications until entering an area of Wi-Fi connectivity. A similar example—also from the mobile operating system domain—is a mobile operating system's preference to perform large software downloads and updates during times of minimal use, allowing the device's operating system to trade processor load (and therefore power consumption and battery life) against the immediacy of software updates.

An entire economics of data and networking optimization exists in which power, time, physical and logical resources, and true financial costs are balanced to optimize a network design for a given user or users or specific application and use case. This is even more pronounced in networks with high mobility and resource constrained hardware in which size, weight, and power (the so-called SWAP trade) are increasingly limited and therefore valuable within a given use case or architecture.

3.1.4 Application Behavior

Application behaviors cause data delays in a network that are not, in and of themselves, related to the availability of links. These delays can be caused by either a direct processing delay by the application (implementation delay) or by the application waiting for some third-party service prior to forwarding the data (dependent delay). For example, if an application needs to log data to a file prior to forwarding that data, then the processing of the data will be delayed—perhaps significantly—based on the availability of the file system and how the software and algorithms are implemented. Alternatively, an application may need to verify an integrity signature on some data prior to forwarding it as part of the security configuration of a network. This verification may require that the application interact with a third-party certificate authority and/or retrieve a public key from such an authority. In this case, the processing of the data in the network is dependent on a third-party and may be delayed by the speed at which a certificate authority can be found. Examples of these types of application delays can also be seen when connecting to HTTPS websites, where multiple rounds of TLS authentications can cause multiple seconds of delay for end users and increases to infrastructure costs and energy consumption [11].

3.2 Terrestrial Internet Approaches to Challenged Networking Environments

Several practical approaches have emerged to avoid or lessen the impact of challenged environments on the terrestrial internet (see Table 3.2). It can be surprising to realize that the efficiency and effectiveness of modern networks remains the result of active, sustained maintenance and optimization efforts rather than a static implementation of well-defined protocols and tools [12]. Evolving approaches are required to prevent growing data volumes from overwhelming existing infrastructure. In all cases, these approaches understand that user data volumes are outpacing the ability to simply replicate readily available resources (compute power,

Table 3.2 Best Practices to Maintain Low-Delay Networks

Practice	Description
Place information	Wherever possible, data in a network should be strategically placed for future consumption. Well-placed data can be faster to retrieve by users and represent a smaller networking load on data providers.
Push data	Data providers can send data when it is generated rather than forcing multiple users to poll for whether data is available thereby decreasing the need for multiple requests for information to be generated by the end users for the same source data.
Ignore state	The less information that must be maintained at the endpoints of a communications network, the faster the data can recover from error and minor issues like out-of-order data delivery.

memory, power infrastructure) and many resources will remain limited regardless of investment (wiring infrastructure, transmission rate limitations, and individual link throughput). This section provides an overview of several approaches to solving networking challenges on the terrestrial internet that do not rely on the brute-force addition of networking resources and that scale with increasing data volumes.

3.2.1 Place Information

Strategic placement of information in a network can reduce the number of network hops required to move data from its source to a data consuming application (see Figure 3.1). By moving data closer to multiple users and further from single producers, the overall number of messages in a network—and the average user wait for data—can be reduced. In a sense this model is like prepositioning of emergency supplies in forward locations where they can be expected to be accessed and drawn upon easily and quickly in time of need. When users are fewer hops to their data, request and respond messages go through fewer waypoint nodes in a network, exist for a shorter period of time in the network, and based on network architecture, may never be communicated outside of a particular subnetwork or enclave.

Electronic mail delivery, as illustrated in Figure 3.1, provides an example of the benefits of this type of information placement. In this figure, solid lines represent network hops that an e-mail message must take when queried by the recipient of the

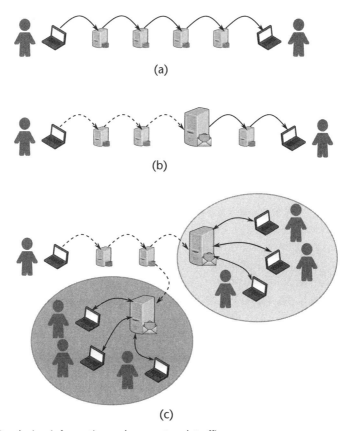

Figure 3.1 Preplacing information reduces network traffic.

e-mail. If information were never preplaced, then every time a user wanted to receive e-mail, they would need to query it directly from the machine that created the e-mail, a situation shown in Figure 3.1(a). This is clearly not a scalable approach. The addition of an e-mail server provides a common place to collect e-mails, such as inside of a corporate intranet. Figure 3.1(b) illustrates this architecture, where dashed lines represent the network hops taken to get an e-mail to the cache, and solid lines continue to represent the hops taken by an e-mail once it is requested by its recipient. Finally, adding multiple e-mail servers can create local areas where e-mail exchange can happen rapidly among users in the same domain while also load balancing e-mail traffic from outside of the enclave. Figure 3.1(c) illustrates this concept where bidirectional arrows represent e-mail exchange within an enclave.

In networks where some nodes have more resources (transmission power, storage, computational power, etc.) than others, well-resourced nodes may be designated as natural data collection points for information generated by less-resourced nodes. A home-based example of this approach is the domain name server (DNS) cache on a home wireless router. Computers within the home typically use the home's router as a default gateway and allow the router to store cached DNS entries. As the node directly connected to, and managed by, the home's ISP node is uniquely resourced to provide this information, to include updating the DNS default gateway provided to the home by the ISP. Another example of this approach farther from the home is a relay spacecraft collecting information from planetary landers so that those landers do not need to transmit data directly back to Earth. In this example, the work of getting data to a common place (from a lander to an orbiter) requires fewer resources than moving data to the end user (from a lander back to Earth).

Different data may be placed in different locations in cases where processing only occurs in a defined region such as a subnetwork, enclave, or geographic boundary. For example, companies may synchronize corporate databases by region, scheduling bulk updates and synchronizations at off-hour times (such as nights and weekends) for that region without needing to centralize the data to do so. In some cases, these regions may be ephemeral or changing, such as is the case with recovery efforts from natural disasters. When a natural disaster requires extended, significant recovery efforts, first responders and other aid workers from distributed geographical regions may converge on a single area. That single area (which may be called a staging location, depot, or command post) provides a common, more heavily resourced location from which to coordinate and deploy resources out into the areas that need them the most and are most resource constrained. Information such as authentication keys for communication equipment can be placed local to the effort, reducing the complexity of coordinating configuration of multiple types of communication equipment. Locating similar data together allows user applications to process only relevant data, simplify configurations, spend fewer resources performing updates, and schedule content changes in ways that are least disruptive to a specific set of users [13].

Data placement strategies do not require data to only exist in one place at one time. Replicating data in multiple network locations can provide distributed users with more timely access to data and less resource burden on producers. The path from an average user to a copy of data can be shortened, requiring fewer hops and less network traffic. Alternatively, by having a large user community access multiple copies of data from multiple sources, instead of a single server, the networking load on that server is reduced. This practice is particularly effective when a relatively stable set of information needs to be read by a large and distributed user base, such as the case with social media, photo-sharing services, video streaming services, and news. DNS, Certificate Authorities (CAs), and regular content aggregators all replace the concept of source-to-destination data exchange with the concept of nearby cache-to-destination data exchange. In all cases, the overhead of duplicative data in the network is insignificant relative to the shortened wait times for users, decreased load on the servers providing these data, and overall reduction in cross-network traffic.

3.2.2 Push Data

There are two ways for a networking application to receive information: it can send a request for that information into the network and wait for the requested data to be sent (the pull method) or it can wait for some source, unprompted, to send information to it (the push method). Examples of the pull method include visiting a webpage or clicking refresh on an application to have it download new information. Examples of the push method include receiving a text message or receiving a notification on many smartphone applications. Determining which method to use is a function of the nature of the data being communicated and the characteristics of the applications operating on that data.

When pulling information, a data consumer determines (a) that it needs a certain set of data and (b) what producers in the network can produce and transmit those data. A data request is sent from the consumer to the producer, and the producer receives and processes the request and transmits the appropriate data back to the consumer. This dissemination method works well when data needs are unpredictable because, in such a case, a push method would not be able to understand what data should be pushed to whom and when. Attempting to push all data all the time is neither efficient nor scalable in any practical networking scenario.

Pull methods work by making several assumptions on the nature of the network, the data, and the data producer. Most importantly, this method assumes that a consuming application can identify the data that it needs, identify the producer of that data, and understands how to request that data from that producer. If there are multiple producers, or producers change over time, the consuming application must manage this complexity. Another assumption is that the network can respond to data requests rapidly because pulled data requires a round-trip data exchange every time any new data is needed. Finally, this method assumes that the application requesting data is the application to receive the data and that the endpoints of the request are both active in the network for the duration of the request and response.

When pushing information, a data subscriber registers a need with one or more data publishers. Those publishers autonomously determine when they have data of interest to their subscribers and then send that data. This dissemination method works well when data needs are well understood in advance and those needs are not so numerous as to overwhelm the computational capabilities of publishers. Latency is reduced because there is no need for the subscriber to make an explicit data request. Network utilization is reduced because there are no request messages in the network and data messages are only present when they need to be. The push method can also be optimized to distribute data via multicast methods in cases where multiple subscribers wish to receive the same information. Receiving data becomes easier as well because there is no need to correlate pushed data with a specific data request; data messages from a publisher are self-contained and self-identifying (for example, data may be timestamped with the time when the data was sampled).

Push methods trade an increase in processing needs on the publisher to decrease the demand for network resources (see Figure 3.2). To analyze who needs what data and when, publishers require significant storage and computation as a function of the timing fidelity of data, the volume of data, the number of identifiable pieces of data, and the overall number of subscribers in the network. While this eliminates the overall number of messages in the network, it is because publishers centralize all data dissemination. If a subscriber is unknown to a publisher, it will not receive any information (unless a multicast protocol is used). Therefore, even when data needs are predictable, a balance must be achieved between the desire to reduce overall network traffic and reducing the computing load on a publisher (see Table 3.3).

In challenged networking environments, preserving networking resources is valued over per-application computation as a practical consideration; networking resources are challenged in these scenarios in ways that computing resources might not be. Even when computing resources are also challenged, applications can work off-line to process data (or restrict overall data volumes) but they have very little ability to affect the underlying network performance.

3.2.3 Avoid Sessions

A session in a computer network refers to the set of information known to two applications so that they can exchange data with certain guarantees, such as ordered,

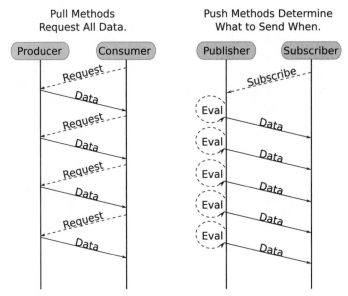

Figure 3.2 Pushing data reduces network traffic.

Table 3.3 The Relative Benefits of Data Dissemination Methods

	Benefits	Consequences
Pull Method	Does not require predictable data needs. Applications manage their own request schedules. Less burden on data providing nodes.	Requires round-trip data exchange. Constantly repeating requests across the network. Requires that an application can exist in the network long enough to receive a reply. Request and response coupled. Data must be identified to be requested. Data consumer must know who produces what data.
Push Method	Unidirectional data path with less propagation latency. Fewer messages in the network. No need to correlate data with a request. Supports multicast operations.	Publishers must know who needs what data and when. Requires data needs to be known in advance. Cannot push data to unknown subscribers.

reliable, and secure delivery. When a message source and a message destination do not need to presynchronize information, the exchange is considered sessionless. Sessions can be used at the network layer, the application layer, or both. At the network layer, sessions help coordinate communications, such as network security, transmission schedules, timeouts, and retransmissions. At the application layer, sessions coordinate semantic information associated with the data exchange, such as the validity of information, application security, timing, and types of data to send next. This section focuses on the difficulty of maintaining network layer sessions and these difficulties can trivially be adapted to application layer sessions.

A session typically has three phases in its life cycle: establishment, maintenance, and teardown. Session establishment involves one or more round-trip data exchanges, where endpoints exchange a variety of information such as security information and link characteristics. One example of such establishment is a TCP session secured with Transport Layer Security (TLS) (Figure 3.3). First, endpoints must exchange information to establish a TCP session, then security information must be exchanged prior to an application being able to communicate data. Session maintenance involves endpoints in the network exchanging context updates, such as timing information, new security keys, and incrementing sequence numbers. Session teardown involves end points deleting their local cache of synchronized information either by mutual agreement or unilaterally (such as when inferring a communications error). Sessions can make bulk message exchange between two endpoints in a network more efficient and more secure, at the expense of node, application, and network resources. There are some cases, however, where the use of sessions is unnecessary and can be safely avoided.

Sessionless data exchange is a simpler solution because there is no requirement for an endpoint to identify, establish, and maintain session state. Sessions can be avoided when the benefits of coordinating messaging endpoints are not necessary for a given message or when coordinated information can be transmitted

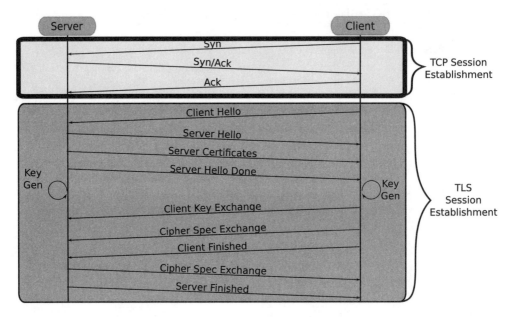

Figure 3.3 The session lifecycle of TCP and TCP+TLS establishment.

with the message itself rather than synchronized separately across the network. There are several cases where sessions are not necessary for data exchange and can be avoided without any impact (and in some cases improvement to) data exchange.

Sessions should be avoided if the data being sent is easy to replace. Often, the decision to use sessions is based on the characteristics of the messages being sent. If messages contain important or otherwise high-priority information sessions may be needed to keep endpoints synchronized and focused on transmission reliability. Less important messages may be sent best effort relying on the network itself for delivery without separate oversight by the messaging endpoints. The determination of which messages are less important can be a function of service agreements or data uniqueness. If the contents of a message are easily replaced or regenerated then the message may be considered less important even if the set of messages, in total, are very important. For example, the use of TCP sessions for online computer games can introduce significant lag because delivering every message in real-time and in order degrades the average delivery time of all messages [14].

Sessions should be avoided if the endpoints of messaging can change in the network, as occurs when using logical endpoints whose associated with physical nodes can change over the course of the message exchange. One scenario that benefits from logical endpoints is the use of multicast addresses to identify interchangeable mobile nodes in a sensor network. Sensors in such a network can be programmed to broadcast their information to whichever mobile device (UAV, spacecraft, car, etc.) is passing by, even if different from the last mobile device used to collect data. Any state information kept on one device (e.g., a UAV) would not help with data exchange to another device (e.g., a passing car) and sessions would need to be destroyed and reestablished every time a new mobile node was used.

Another scenario that benefits from logical addressing is the use of load-balancing and/or anonymizing servers. These servers determine which physical nodes receive traffic as a function of the server's configuration and purpose. This is a popular approach when handling requests from a large user base, such as the case with web search engines or social media sites. In these cases, requests can go to one of a small number of servers which then pass requests on to other servers to perform server-side processing. Clients believe they are communicating with a single server when, in reality, their requests are passed to multiple physical servers, in some cases with the specific physical server changing between each received message.

Sessionless data exchange in these environments handle endpoint substitutions more effectively because they do not require all potential endpoints to share the same state information.

3.3 Terrestrial Internet Design Patterns

An internet design pattern refers to a protocol or algorithm design whose implementation exemplifies the best practices referenced in the prior section. Four patterns are described in this section: content distribution, publish-subscribe mechanisms, autonomic computing, and stateless data exchange (see Table 3.4).

3.3.1 Content Delivery Networks

A content delivery network (CDN) [15–17] is one in which geographically distributed servers are prepopulated with content that is likely needed by users within a geographic region. CDNs are typically overlays on the terrestrial internet and cache traffic for users. When the cache is located physically closer to an end user there is a higher likelihood that retrieving the data takes fewer network hops, has lower latency, and that retransmissions will use less network resources (throughput and energy). When considered as part of network architecture, caching nodes can be purposely built and resourced, including additional storage, network connectivity,

Table 3.4 Design Patterns Follow Best Practices for Constrained Data Exchange

	Place Information	Push Data	Avoid Sessions
Content delivery networks	Caches places strategically in a network.	Data producers push data to their cache locations.	No state maintained between a data producer and a data consumer.
Data subscriptions	Brokers can be used to centralize subscriptions and published data.	Publishers send data to subscribers.	No state maintained between a publisher and a subscriber.
Autonomic computing	Autonomy acts on local information.	Rules and automation for sending information.	No closed-loop control.
Stateless data	Local information takes the place of negotiated, synchronized state.	Data exchange initiated by either producer or consumer.	Sessions unnecessary for stateless data.

and computational resources as appropriate. The terrestrial internet would not function as we are familiar with it without using the architectural approach of distributed content caches. This does not mean that the concept of local caching is without issues or obstacles to effective implementation.

There are three significant issues with content caches: where to place them, how large they should be, and how long data should exist in each cache. Selecting cache locations is a complex function of considerations such as the cost of energy, government regulations, annual maintenance, political factors, and number of end users serviced by the cache. Simple deployments can be placed as a function of the number of local users, projected user needs, historical data flow analysis, and real-time measurement of requests. Caches in a network have different capacities. Well-resourced nodes that access reliable power and communications infrastructure may have terabytes or petabytes of storage and their own local caches to quickly serve common data. Less resourced nodes might not be able to commit similar resources in terms of storage and processing capacity as a function of their supported data rates. Regardless of cache size, over time a cache will fill unless there is some strategy for expiring information or otherwise limiting what new information is allowed into the cache.

There are many familiar examples of this CDN pattern on the terrestrial internet (see Figure 3.4). Content aggregators prepopulate data caches regionally. Data owners, such as audio and video streaming services, distribute servers globally to reduce video and audio delays for users and server loads. The same patterns are repeated for cloud data centers. Most of the terrestrial internet, as we experience it today, is not a model of end-to-end data exchange. Instead, it is the storage of long-lived, multiuser data in a series of strategically placed caches.

For the terrestrial internet, end-to-end data exchange is not preferred because of the need to scale data volumes, increased user communities, and the impracticality of overengineering networks to always carry worst-case traffic volumes. To the extent that data is stored in a cache, data exchange is no longer end-to-end from data producer to data consumer. Instead, there are a series of data exchanges; some populating the caches, and some reading from the caches.

In a challenged networking environment, a cache miss might be a significant event, as retrieving the data from an original source or some up-stream cache, might incur very long delays or otherwise consume nontrivial amounts of resources compared to those available. This means that cache sizing and cache expiration strategies must be carefully considered. Nodes in a challenged networking environment exist without access to reserves of networking capacity such as processor or storage capacity, and, potentially extending to lack of regular power, connectivity, or other hard limiting factors. These nodes may be unable to carry a large cache or unable to maintain the data rates necessary to effectively populate or keep such a cache updated. Often environmental challenges occur because it is not possible to densely populate an area with networking resources and node mobility may change the best place to store cached data. As such, it can be difficult or impossible to place caching nodes. One way to overcome uncertainty on cache placement is to make every node in the network a caching node. This solves the general problem of cache placement but requires new protocols and algorithms to handle the storage of data throughout the entire network.

3.3 Terrestrial Internet Design Patterns

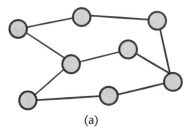

(a)

Without caches, data is assumed to transmit end-to-end without delays or disruptions, shown in solid lines

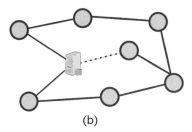

(b)

Adding caches in the network allows delayed reception of data that was produced earlier. Delayed links are shown with dashed lines.

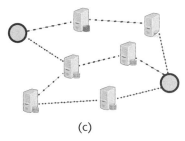

(c)

When every node can serve as a cache, all links can be delayed and network challenges more easily overcome. But this approach requires changing the protocols and algorithms used to implement CDNs today.

Figure 3.4 CDNs maintain caches using high-reliability links.

3.3.2 Data Subscriptions

In a publish-subscribe (PubSub) pattern, an application (the subscriber) requests sets of named data that it would like to receive when those data sets are created by other nodes (the publishers) in the network. The request (a subscription) is sent by the subscriber to one or more PubSub points of contact (the managers) who then ensure that the subscriber receives the data they desire as it is published. The general data flow through a PubSub pattern is illustrated in Figure 3.5. There are two ways in which a subscription can be tracked and implemented in this pattern: with the manager handling all subscriptions or with the manager delegating subscriptions to the publishers that are affected by them.

Figure 3.5(a) illustrates the approach of centralizing all subscription processing within the manager itself. In this approach subscribers only ever communicate with the manager and the number of publishers in the network is effectively abstracted away. Similarly, publishers can send their data directly to managers without understanding the number of subscribers that exist in the network. There are several benefits to centralizing subscription processing. Managers can deduplicate or otherwise fuse data prior to sending it to subscribers, which may be impossible were publishers to communicate directly with subscribers themselves. Also, both publishers and subscribers only need to authenticate to the manager and otherwise never need to communicate directly with each other which can increase the operational security within the network. The significant downside to this approach is that it places a significant burden on the manager (and on the network throughput into and out of the manager) to process large and potentially high-rate data volumes across potentially large numbers of publishers and subscribers.

Figure 3.5(b) illustrates the approach of decentralizing subscription handling by having managers configure publishers to directly communicate with subscribers. This approach scales with the number of subscribers and publishers in the system, and the manager is only used to configure publishers with their unique set

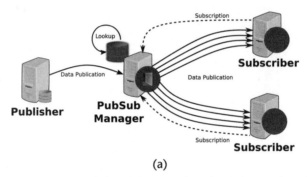

(a)

A manager can be a single point of contact for subscriptions and produced data

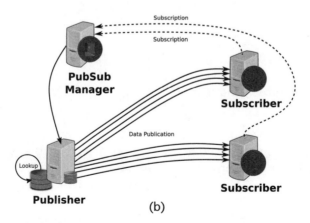

(b)

A manager can delegate subscriptions to publishers to reduce bottlenecks in a system

Figure 3.5 Centralized and distributed models in PubSub architectures.

of subscribers. By load balancing across the whole set of publishers in the network, the manager itself does not become a performance bottleneck in the system.

In any subscription-based system there must be some data identification scheme and this scheme must balance giving subscribers flexibility and not overwhelming publishers tracking what data goes where. For example, allowing subscribers to track every datum generated by a publisher gives maximum flexibility but requires the publisher (or manager) to map every datum to every subscription. If publishers generate millions of data values supporting thousands of subscribers, this mapping can be resource intensive. An alternative scheme is to let publishers define sets of data that can be subscribed to. By only subscribing to curated data sets, the complexity of the subscription mapping is reduced at the expense of some loss of fidelity at the subscriber who may receive more data than necessary. Defining the scope of a data subscription requires an understanding of the computational power of publishers, managers, and subscribers, which data in the system must be used together, and the ability of the network to tolerate larger or more frequent messages.

There are several benefits to using a PubSub pattern in challenged networking environments. Subscription models do not require that publishers and subscribers be active in the network at the same time. Data subscriptions persist even if a subscriber is temporarily disconnected from the network. For example, network monitoring applications may monitor User Datagram Protocol (UDP) ports for status updates and react to them as they are received. If the application is stopped and restarted, the newly restarted application may still receive pushed network data without having to rerequest a data subscription. In addition to allowing subscribers to recover without needing to restart a subscription, a publisher can continue to send data to a subscriber even when the subscriber is offline. In store-and-forward networks, this pushed information can be help in the network until the subscriber is brought online later.

Subscription models do not require sessions to operate and do not rely on synchronized state to function. Published data is named, uniquely discernable (such as using timestamps or a nonce), and otherwise annotated with any necessary context information. Issues such as out-of-order delivery and missed data can be resolved by evaluating the information included in a published data set rather than requiring sessions to maintain this information or a sequence of additional query/response cycles to provide clarity.

Subscription models do not require excess bandwidth. Publishers only generate data when necessary. In pull systems, consumers must query data without knowing whether the data has changed resulting in network activity even in cases where there is not new data to be communicated. As the number of consumers and data items grow, pull systems can congest networks even in cases where there is no data to be pulled.

3.3.3 Autonomic Computing

Autonomic computing refers to an approach for distributed applications to manage themselves using local sensing and configuration to hide the complexity of distributed management from end users [18]. Actions can be either automated or autonomous based on the freedom given to the responding application. Automated actions are fixed sequences that have been preplanned for the encountered event. Autonomous actions are customized for a specific event and may include the parameterization of existing commands and the self-selection and recombination of commands or command sequences.

A deep-space spacecraft serves as an excellent example of a platform that requires both automated and autonomous actions. Such a spacecraft may be light-minutes to light-hours from Earth, preventing operators from correcting errors in real time and requiring some level of autonomic computing. Consider a flight software application that manages a set of redundant heaters used to keep its spacecraft operating within a specified temperature range. When the activity of the spacecraft changes such that the internal temperature rises or falls (such as turning on flight computers or instruments) then the flight software application will turn on/off heaters to keep the temperature steady. If a heater breaks, the application can sense that scenario and start using a backup heater instead. The application performs an autonomous action when it determines the desired heater configuration for the spacecraft—the desired temperature setting is not known in advance and is calculated by the application as a function of its local state. The application performs an automated action when it switches out a failed heater for its designated backup heater—the replacement of a failed component was programmed in advance as a known response to a predicted event.

One design for autonomic computing is the expression of events using predicate logic statements and the implementation of a predicate logic analyzer to determine whether any initiating events have occurred and what types of actions should be taken in response to those events. Predicate expressions can be preconfigured, asserted by some network management function, or self-determined as a function of the application's other autonomous actions. Response actions can be assigned to these predicates as part of the configuration of the system.

Both CDN and data subscription patterns require the existence of autonomic applications for their proper function. CDNs must determine what data to cache, in which cache(s) to place it, and how long that data needs to stay in the cache. Data subscriptions must provide enough fidelity in their subscription model to balance overwhelming subscribers with data and overwhelming publishers with computation. Separate from the issue of data identification, some Subscribers may only want data values pushed in certain circumstances as a function of the state of the producer. In this situation, the producer can be configured with an automated action (always send data when a value changes) or with an autonomous action (determine whether a subscriber would want to know about a new value). Both initiating events can be captured as predicates (see Figure 3.6).

Two examples of applications used on the terrestrial internet that apply this type of autonomic architecture are Simple Network Management Protocol (SNMP) [19] traps and If This, Then That (IFTTT) automation [20]. The SNMP is used to configure and monitor devices over a network. It is considered a simple protocol because it consists of three message types: GET, SET, and TRAP. GET messages are used to retrieve a data value from a managed device, whereas SET messages are used to set a value on the device. For both GET and SET messages, a managing device determines a need to monitor or change a configuration and sends the appropriate message to the managed device. In the case of a TRAP, the managed

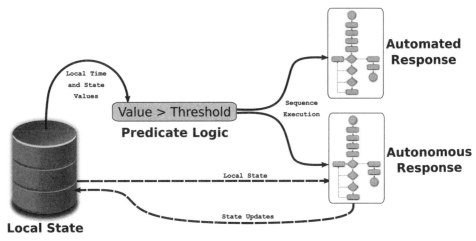

Figure 3.6 Predicated automated and autonomous responses.

device itself determines that it must send information to the managing device. A common use case for TRAP messages is when a managed device experiences some significant error condition and must send an alarm or alert message to the managing device. SNMP TRAPs are examples of automated responses.

IFTTT refers to a family of applications creating if-then statements that coordinate behavior across multiple web-based services. As new web-services become integrated into the IFTTT system, their status can be used to form predicates that, when true, result in some user-defined action. The IFTTT predicates were originally termed recipes and later were renamed applets. Examples of the types of service-based autonomic behavior enabled by IFTTT include sending an e-mail when a hashtag is posted on a social media feed or turning on a smart lightbulb at a certain time.

The complexity of implementing methods such as predicate logic analyzers must be evaluated versus sending potentially unneeded data to subscribers. Often the trade becomes one of processing resources available to the publishing node versus networking resources available for transmitting data to the subscriber nodes. When operating in a challenged network, networking resources may be significantly limited relative to individual node processing power. As such, it is often the case that finer grained data identification and predicates be used to help identify when a subscriber would, and would not, be interested in receiving a data notification.

While fully autonomous network management systems are emerging in both research and engineering communities, they are problematic in challenged networking environments for a variety of reasons. First, fully autonomous systems need proper training data to establish baseline behavior and must be able to review the effects of actions to see if responses need to be tuned. Challenged environments rarely provide this type of closed-loop control necessary to keep fully autonomous systems working correctly. Second, when nodes in a network cannot communicate regularly with each other, then their coordination must be prescribed through a more deterministic method – for example a message could be sent to all applications in a challenged networking environment to perform an action at exactly 2 p.m. So long as all messages are received prior to that time, the desired behavior can be coordinated. If a node autonomously changes its state prior to that time, and without the ability to inform other nodes of its change, then the desired coordinate response may fail to occur. Finally, some networks must retain the information necessary to forensically reconstruct why a causal chain occurred as part of error analysis. This is particularly important in cases where an application in a challenged networking environment is found to be in an error state and is only seen periodically by other management nodes. In this case, trying to understand how a node came to be in an error state is difficult, and at times, impossible in a fully autonomous system.

3.3.4 Stateless Data

Maintaining session state between two messaging endpoints requires computing and networking resources. Networking bandwidth must be allocated for the exchange of session establishment and synchronization. Endpoints must compute the necessary state to be synchronized and react to changes in the overall session and the variety of errors that can occur as part of the establishment and synchronization process. In individual cases over high-rate, high-availability networks, this establishment and maintenance occurs in milliseconds and rarely creates a significant problem for endpoints or the network. However, as the number of discrete requests requiring discrete sessions increases, common endpoints could be faced with the possibility of establishing hundreds or thousands of individual sessions.

In a client-server networking model, multiple clients may request information from a single server resulting in a small number of exchanges with each individual client. When client requests are stand-alone, establishing a unique session with each client incurs the work of session establishment without the benefits of then using that session to make subsequent messaging go faster. When clients only exchange a small number of messages with a server, patterns for sessionless data exchange have evolved out of necessity. This scenario happens regularly with web servers connecting with terrestrial internet browsers acting as clients. When browser requests are stand-alone, they do not need to occur within the context of a session.

An example of this type of sessionless state exchange is the Representational State Transfer (REST) [21] architecture in which a server publishes a set of services that can be called by clients without any client-specific state being kept on the server itself. Important to this concept is that every client request contain all information necessary for the server to process the request. RESTful architectures enforce other best practices, such as uniform interfaces and separation of concerns between clients and servers. However, the lack of per-client state that allows the system to scale to hundreds and thousands of clients without overwhelming webservers.

This is not to imply that web-based traffic is itself stateless. One of the most popular uses of RESTful interfaces is the Hypertext Transfer Protocol (HTTP) which has no built-in mechanisms for in-order message assembly or retransmission. HTTP works best when running over a reliable networking layer that can perform this level of error recovery and the most common transport for HTTP messages is the TCP, which does establish a networking session between the client and the server in the RESTful transaction. However, to avoid even this level of session-based exchange, the newest version of HTTP (HTTP/3) has been designed to use Quick UDP Internet Connections (QUIC) [22] as the underlying transport mechanism. QUIC provides a level of message ordering and retransmission without requiring the same resource burden of TCP and without requiring the multiple handshakes needed to establish a TCP connection.

Stateless application data is needed in DTN-like scenarios because there is never a guarantee that the endpoints of a messaging exchange are active in the network at the same time. If one endpoint leaves the network, then there can be no state synchronization with that endpoint while it is away from the network. If the two endpoints both come and go in the network over time, they may never coexist long enough to perform meaningful synchronization. Like the problem encountered by webservers, attempting to maintain some state on a common endpoint is unscalable as the number of connections increases. While this is certainly an issue on the terrestrial internet where thousands of connections can be received each second, it is also an issue in more constrained networks where the number of connections may be far fewer, but session state may need to be held for extremely long periods of time (minutes, hours, or days).

3.4 Summary

As the most successful computer network in the history of humanity, the terrestrial internet defines our concept of how people and machines should communicate. An exhaustive review of technologies that enable this system to continue to operate is well beyond the scope of any single publication. However, there are conditions where data on the terrestrial internet experience delays and disruption like those experienced in a challenged networking environment, particularly where users wish to have common networking features in areas that do not have common networking infrastructure. The novel architectures and applications that keep terrestrial internet traffic flowing provide examples of how similar architectures and applications addressing the limitations of challenged networks in general may be built with similar beneficial effect.

Each of these novel architectures and applications share a common concern: the network is going to run out of resources. Content delivery assumes that the network, or the content provider, will run out of resources to serve data without providing load-balancing caches. Data subscriptions assume that round-trip communications will overwhelm network traffic at scale. Autonomic computing assumes that operators will not be able to communicate with network nodes in a timely manner to fix issues. Stateless data assumes that nodes in the network will run out of individual processing resources and networking resources to keep information synchronized.

To build delay-tolerant applications, developers must maintain an understanding of where the resource limitations will emerge in the operation of the network. On the terrestrial internet, challenges can be addressed by assuming that lack of resources in one part of the network can be overcome by extra resources in other parts of the network. For example, a RESTful interface may not use sessions because those resources are not available at the application layer, but those stateless requests may be communicated over a TCP session assuming state information is available at the networking layer.

As increasingly challenging environments are networked, resource limitations will emerge in every part of the network and the solutions that work on the terrestrial internet will no longer be sufficient. Fundamentally new protocols, algorithms, and architectures are needed, and in fact are being developed, by the networking research community.

3.5 Problems

3.1 Consider a subscription-based service allowing users to view movies on a variety of web-enabled devices. Provide an example of why such a service would experience each of the four types of DTN-like problems even when users connect to the service over the terrestrial internet.

3.2 Assume that you are a network architect and must determine what nodes in the network should serve as content caches of information. List three different criteria that could be used to assess whether a particular node would make a good content cache.

3.3 Describe a real-world example of a push service and provide three reasons why such a service would not function as a pull service.

3.4 Assume that you are attempting to establish a TCP+TLS session across a four-hop medium-Earth orbit (MEO) satellite network and a single hop to ground. Each satellite hop incurs a 17-ms signal propagation delay and a 7-ms delay for processing, per satellite. The ground hop terminates in a ground station and has a 100-ms signal propagation delay and another 4 ms of processing delay. Assuming all other terrestrial internet traffic happens instantaneously, and message size is not a factor, how many milliseconds would it take to establish a TCP+TLS session in this environment?

3.5 Provide four different content expiration functions that could be used to expire data from a content cache. Explain in what scenarios each function would be most appropriate.

3.6 Consider a set of 200 data items generated by a producer every second and 100 subscribers. If data are identified individually, and each subscriber has an individual subscription to each data item, how many subscriptions must a producer process every second? What is the maximum time a producer must spend on each subscription each second?

3.7 Describe in your own words the difference between an automated and an autonomous action. Provide a real-world example of each.

3.8 Describe a scenario where attempting to synchronize state information across a network results in a failure to communicate. What conditions would need to be true of the network itself for this to occur?

References

[1] Maher III, R. D., et al. "Method and Apparatus for Enforcing Service Level Agreements," U.S. Patent No. 7,272,115, September 18, 2007.

[2] Marcon, M., et al. "The Local and Global Effects of Traffic Shaping in the Internet," *2011 Third International Conference on Communication Systems and Networks (COMSNETS 2011)*, IEEE, 2011.

[3] Georgiadis, L., et al. "Efficient Network QoS Provisioning Based on Per Node Traffic Shaping," *IEEE/ACM Transactions on Networking*, Vol. 4, No. 4, 1996, pp. 482–501.

[4] Cheng, H. K., S. Bandyopadhyay, and H. Guo, "The Debate on Net Neutrality: A Policy Perspective," *Information Systems Research*, Vol. 22, No. 1, 2011, pp. 60–82.

[5] Krämer, J., L. Wiewiorra, and C. Weinhardt, "Net Neutrality: A Progress Report," Telecommunications Policy, Vol. 37, No. 9, 2013, pp. 794–813.

[6] Mitchell, H. B., *Multi-Sensor Data Fusion: An Introduction*, Springer Science & Business Media, 2007.

[7] Kim, M. W., et al. "Battery Lifetime Extension Method Using Selective Data Reception on Smartphone," *The International Conference on Information Network 2012*, IEEE, 2012.

[8] Hail, M. A., et al. "Caching in Named Data Networking for the Wireless Internet of Things," *2015 International Conference on Recent Advances in Internet of Things (RIoT)*, IEEE, 2015.

[9] MacKie-Mason, J. K., and H. R. Varian, "Pricing Congestible Network Resources," *IEEE Journal on Selected Areas in Communications*, Vol. 13, No. 7, 1995, pp. 1141–1149.

[10] Lee, J., et al., "Economics of Wi-Fi offloading: Trading delay for cellular capacity," *IEEE Transactions on Wireless Communications*, Vol. 13, No. 3, 2014, pp. 1540–1554.

[11] Naylor, D., et al., "The Cost of the S in HTTPS," *Proceedings of the 10th ACM International on Conference on emerging Networking Experiments and Technologies*, ACM, 2014.

[12] Handley, M., "Why the Internet Only Just Works," *BT Technology Journal*, Vol. 24, No. 3, 2006, pp. 119–129.

[13] Betts, B. J., et al., "Improving Situational Awareness for First Responders via Mobile Computing," *NASA Techinical Reports*, 2005.

[14] Chen, K.-T., et al., "An Empirical Evaluation of TCP Performance in Online Games," *Proceedings of the 2006 ACM SIGCHI International Conference on Advances in Computer Entertainment Technology*, ACM, 2006.

[15] Pallis, G., and A. Vakali, "Insight and Perspectives for Content Delivery Networks," *Communications of the ACM*, Vol. 49, No. 1, 2006, pp. 101–106.

[16] Vakali, A., and G. Pallis, "Content Delivery Networks: Status and Trends," *IEEE Internet Computing*, Vol. 7, No. 6, 2003, pp. 68–74.

[17] Buyya, R., M. Pathan, and A. Vakali (eds.), *Content Delivery Networks*, Volume 9, Springer Science & Business Media, 2008.

[18] Kephart, J. O., and D. M. Chess, "The Vision of Autonomic Computing," Computer, Vol. 1, 2003, pp. 41–50.

[19] Feit, S. M., SNMP: A Guide to Network Management, McGraw-Hill Professional, 1993.

[20] Ovadia, S., "Automate the Internet with "If This Then That" (IFTTT)," Behavioral & Social Sciences Librarian, Vol. 33, No. 4, 2014, pp. 208–211.

[21] Fielding, R. T., and R. N. Taylor, "Principled Design of the Modern Web Architecture," *ACM Transactions on Internet Technology*, Vol. 2.2, 2002, pp. 115–150

[22] Gratzer, F., "QUIC-Quick UDP Internet Connections," Future Internet and Innovative Internet Technologies and Mobile Communications, 2016.

CHAPTER 4
Rallying the Research Community: DARPA, NASA, and Disruption Tolerance

Beginning in the 1990s, mobile communications in general—and mobile networking in particular—experienced an explosion of new applications, new technologies, and new operations concepts. Terrestrial communications applications now included cellular telephone and cellular networks, mobile data modems were becoming increasingly common, and the idea of deploying remote monitoring and control systems across widely distributed geographies began to grow within industry, government, and the military communities. In the airborne domain, digital data links were growing in prevalence between aircraft, ground controllers, and remote monitoring systems. In space, the National Aeronautics and Space Administration (NASA) and its international partners began to experiment with deep space relay techniques around Mars. With these new commercial, military, and civilian government demands, research into techniques and architectures to enable reliable communications in a challenged networking environment began in earnest.

This chapter reviews the history of research into DTN as a solution to the problems of communications in challenged networking environments. In doing so, it captures the evolution of DTN from initial observation of new networking modalities to a common set of features and technologies for the challenged networking community. Exploring the evolution of DTN to migrate from resourced network environments to challenged network environments will inform efforts to migrate from resources network applications to delay-tolerant network applications.

4.1 History of Delay-/Disruption-Tolerant Research

Research into solutions for challenged networking environments began in earnest in four different communities (space, defense, academia, industry), each with its own set of technical, operational, and physics-based challenges. Each of these communities discovered their challenges in the context of their own operational needs. One significant branch of study resulted in the body of work that we now commonly refer to as DTNs.

The civilian space agencies were interested in finding ways to more efficiently move spacecraft commands, telemetry, and mission (science) data around the solar system, including to destinations that are not always in line-of-sight with Earth. As the defense community increasingly came to rely on information at all levels—from the strategic commanders down to individual soldiers—it became interested in the potential of high-reliability communication in a mobile, spectrally contested environment. The academic community grew interested in exploring the problems posed by data routing, network establishment, addressing mobility in network topologies, distributed cryptography, and many other areas of mobile networking. Finally, the commercial/industrial sector, while generally following the lead of government and academia, identified the potential markets for systems with DTN-like properties and engaged through several forums including development of in-house proprietary approaches for sale to interested users and participation in the various international networking and communications standards bodies such as the Institute of Electrical and Electronics Engineers (IEEE) and Internet Engineering Task Force (IETF).

Each of these communities worked to create point solutions to solve individual problems for individual missions. Such solutions, by their nature, are difficult to re-apply to other projects or in other communities. If these community-specific problems could be reimagined as a general challenged networking problem, then solutions to that general problem could be shared by all. Several research organizations undertook the task of defining (and addressing) the general networking problem. The remainder of this chapter provides a discussion of how these organizations approached the networking problem, how they worked with each other, and the results of their efforts.

4.2 NASA and DARPA

In 1998, the U.S. Defense Advanced Research Projects Agency (DARPA) provided funding to NASA's Jet Propulsion Laboratory in Pasadena, CA to explore architectures and begin the development of an interplanetary internet and to participate in the development of DTN technologies, architectures, and operations concepts as part of DARPA's Next Generation Internet Initiative. The premise for the Interplanetary Internet was based on the principles of a store-and-forward network routing scheme suitable for use in high-delay, disrupted environments. DARPA's intent in funding DTN work and establishing communities of practice throughout academia and industry was as Leigh Torgerson, the first task manager for the DARPA-funded Interplanetary Internet study stated in an e-mail on the DTNRG list serve, "I can tell you that we very deliberately set up the DTNRG in the IRTF […] and funded the academic community to insure that DTN had as broad an applicability and research gene pool as possible [1]." To accomplish the goal of a broad research gene pool, DARPA funded seed activities at NASA, universities, and industry, eventually awarding Phase I-III activities to BBN Technologies [2].

One of the leading groups that identified the potential benefit of addressing the general networking problem of delay/disrupted networks was the NASA and DARPA supported team of NASA's Jet Propulsion Laboratory, and the MITRE and SPARTA corporations. This team, growing out of the DARPA funded study, and working with internet pioneer Vint Cerf, worked across a variety of standards organizations and communities eventually including the Consultative Committee on

Space Data Systems (CCSDS), IETF/IRTF, U.S. Department of Defense (DoD), and Interagency Operations Advisory Group (IOAG)[1], and sought to include academia, industry, defense, and the space and terrestrial operations communities to serve as a bridge between the various groups building DTN capabilities.

The DARPA funded NASA activities included significant development and testing of DTNs and the supporting protocols for network, link, and physical layer interfaces as well as architecture and operations studies. The NASA-Jet Propulsion Laboratory (JPL) DTN Core Engineering activity included NASA's engagement with the DTN Research Group (DTNRG), authoring and coordination of several internet standards for DTN protocols as RFCs within the Internet Research Task Force (IRTF), and published initial Internet Drafts for the Compressed Bundle Header Encoding (CBHE) [3], Contact Graph Routing (CGR) [4], Bundle Protocol Extended Class of Service (BP ECOS) [5], and the DTN URI Scheme [6]. Additionally, JPL's DARPA funded work included technical development and test of various DTN APIs implemented in the NASA/JPL Interplanetary Overlay Network (ION) DTN Bundle Protocol Agent (BPA), the DTNRG DTN2 implementation, and the commercially developed (also under DARPA sponsorship) BPA from BBN Technologies. JPL also provided several reference implementations of user-level applications including the Simple Data Recorder (SDR) and Asynchronous Messaging System (AMS) to the DTNRG DTN2 API as well as an implementation that has been qualified for space flight applications [7].

In 2010, DARPA conducted a set of large-scale, live exercises demonstrating the applicability of the developed DTN architecture and protocols to the military. These tests consisted of groups of marine vehicles during maneuvers at Marine Corps Base Hawaii and Camp Pendleton and involved the coordination of the USMC, DARPA, SPAWAR, NASA/JPL, MITRE, BBN Technologies, and the Sparta Corporation. These exercises brought together the implemented DTN protocols, network service management, operator interfaces, and real-world mobility and tactical maneuver scenarios. The tests included RF, line of sight, and SATCOM links and represented a real scenario the marines would likely encounter.

Following the DARPA Phase III activities (see Figure 4.1), NASA began funding internal DTN development as a core technology area within the Space Communication Program (later the Space Communication and Navigation Program). NASA's efforts over the following years included further development and refinement of ION (mostly by JPL, NASA's Goddard Space Flight Center, with support from the Johns Hopkins University Applied Physics Laboratory (JHU/APL)) and DTN2 at NASA's Marshall Space Flight Center. NASA envisioned the resource optimized ION implementation for flight applications and DTN2 for terrestrial, ground site, and control center applications. The need to test two independent implementations of the same standard specification for interoperability in part drove NASA to invest in parallel DTN implementations moving the ball towards standardization within the CCSDS.

---

1. The role and history of each of these organizations will be discussed in detail in following sections.

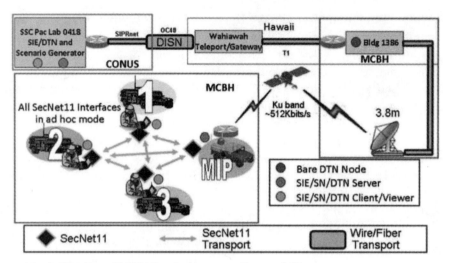

Figure 4.1 DARPA Phase III DTN Test Architecture. (Source: [7].)

4.3 International Space Agencies

As cooperation between international space agencies increased through the 1990s, these agencies realized that it was difficult to interoperate using each other's communications and tracking networks as well as to operate joint missions with multiple spacecraft or systems in an integrated way. NASA and the European Space Agency (ESA) began discussing ways to address these challenges of interoperability. During a 1998 meeting between NASA and ESA, the two agencies determined that issues and challenges faced by NASA and ESA systems and spacecraft interoperating would best be served through a collaborative multiagency forum. This was the birth of the Interoperability Plenary (IOP), consisting of senior managers of the agencies, with the first formal session occurring in June of the following year. The first IOP included leadership from multiple civilian space agencies, including

- NASA, United States;
- ESA, multiple European countries;
- Japan Aerospace Exploration Agency (JAXA), Japan;
- Deutsches Zentrum für Luft- und Raumfahrt e.V. (DLR), Germany;
- Agenzia Spaziale Italiana (ASI), Italy;
- Centre National D'Études Spatiales (CNES), France.

The stated goals of the IOP, agreed to by the six agencies at the first IOP meeting (IOP-1) were

- To reach multiagency agreement on the need for interoperable space communications and navigation architectures;
- Accordingly, member agencies would provide space communications and navigation resources for interoperable cross support to member agency

spaceflight missions, including human, near Earth, lunar, and deep-space missions, as mutually agreed [8].

The structure established by the various government agencies was hierarchical with the Interoperability Plenum consisting of senior leadership directing lower level working bodies. The first meeting of the Interoperability Plenary (IOP-1) established the IOAG as the working body within which specific technical, operational, and policies studies would be conducted to accomplish the objectives of standardization and interoperability. The IOAG in turn, as a management level organization, established and directed the IOAG Space Internetworking Strategy Group (SISG) to execute the detailed trade studies, architecture studies, and technical analyses necessary to develop the recommendations for the IOAG to present for ratification to the IOP.

4.3.1 History of IOP Activities

IOP-1 was held at ESA Headquarters in June, 1999. During this meeting, representatives from international space agencies met to discuss and agree on the need for increased interoperability and the method for its implementation. The IOP discussion focused on four main areas of interest among the agencies: past interoperability mission experiences—both positive and negative—the need for interoperability for each agency's future missions, the specific assets of the agency that could be made interoperable among the international community, and the current status and plans to implement interoperable standards [9].

NASA hosted the second IOP (IOP-2) in December 2008. During this meeting, the member agencies reviewed the progress made by the IOAG in space communication interoperability. A communiqué was issued providing resolutions for guiding the future direction of the IOAG and its related activities. These activities were to be completed over the course of several years, resulting in the presentation to the IOP for ratification during its third meeting. IOP direction to the IOAG included the creation of a draft Solar System Internetwork (SSI) Operations Concept and service catalog and a mature architectural proposal for review and endorsement [10].

The third IOP (IOP-3) was hosted by CNES in June 2013. At this meeting, the IOAG/SISG presented the results of its work for endorsement by the space communication leadership of the member agencies and to decide which direction to take on new activities of the IOAG since IOP-2 [11].

4.3.2 Establishment of the IOAG

As a result of the first IOP meeting in 1999, the IOAG was established in order to determine the best ways to achieve cross support of space communication and navigation infrastructure and capabilities across the international space community and to expand the enabling levels of space communications and navigation interoperability.

The goals of the IOAG as chartered by the IOP are:

- Understand issues related to interagency interoperability and other space communications matters;
- Identify solutions complying with IOP policies;
- Recommend resolutions to the IOP for specific actions by the IOP [12].

Within the context of the IOAG, several working groups were established to develop architecture and operations concepts, networking protocols, test and validation implementations of DTN approaches, and policies for use of interoperability techniques within the various space agencies.

4.3.3 The Space Communications Architecture Working Group

NASA's Space Communications Architecture Working Group (SCAWG) was established as an internal policy and architecture body whose purpose was to establish NASA's space communications and tracking architecture framework through the 2030's. Coordinated by NASA Headquarters, the SCAWG included technical and programmatic members from each major NASA field center, the JPL, and the NASA mission directorates. While a member of the IOAG, NASA established the SCAWG to develop a US perspective on space communication and future architecture both for the agency itself and to inform NASA's interactions with its international partner space agencies in the IOAG and the IOP.

The charter of the SCAWG spanned all aspects of space communication and navigation for NASA's future systems (at the time) and included specific exploration of network, security, spectrum, and navigation architectures. Each of these architectures was applied to operating mission concepts as well as NASA's prediction of future robotic and human exploration in the near-Earth, Cis-Lunar, Martian, and deep space regions. The SCAWG further explored options for terrestrial space communication and navigation (e.g., Ground stations and control centers) as well as communications relay and navigation systems in the near-Earth (Earth orbiting), lunar, and Martian regions of space.

The resulting architecture and operations trades produced a series of recommendations. Most significant in a discussion of DTNs is the concept of space communication of the future being based on an interconnected network of networks model [13]. Among the recommendations of the SCAWG was a set of keys enabling technologies that would be required to make this future realizable. Primary in that set of key technologies was the concept of advanced distributed networking and network technologies (see Figure 4.2) to extend terrestrial networking capabilities into space.

A key recommendation from the SCAWG Final Report was that having identified networking and a distributed network-of-networks topography as the preferred architecture, future research and development was required in the areas of network services and protocol identification and development, identifying and standardizing the policies and mechanisms for governing and operating these services and standards across the networks, control centers, science centers, and flight systems. Important to the evolutionary approach recommended by the SCAWG, the future protocols and governance processes would need to provide the ability to work from current architectures, support the current set of missions, networks,

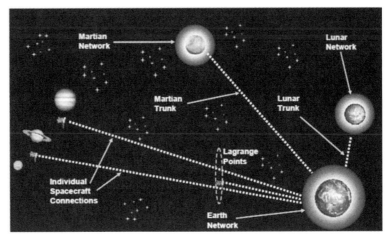

Figure 4.2 SCAWG recommended space communication architecture concept. (Image credit: NASA.)

and mission operations, while providing advanced capabilities to future missions in an evolutionary way.

The driving requirements for a future space communication network, a so-called interplanetary network, were identified as:

- Provide multilevel mission data communication services for legacy missions, new science missions, and new (human) exploration missions;
- Support IP routing and internet applications for space and ground elements;
- Provide data communication service on-ramps for future government, and, potentially, commercial service providers;
- Accommodate both scheduled and unscheduled communications;
- Accommodate both continuous and intermittent connectivity;
- Provide service over space links characterized by both large and small signal propagation latencies, both unidirectionally and bidirectionally, and both low and high BERs;
- Support a wide variety of data types ranging from critical command and telemetry, file transfer (including web pages), messaging (such as IM or e-mail), voice and video streaming;
- Provide certain qualities of data communication services depending on mission requirements including isochrony, reliable transfer, preservation of order of transmission, timeliness, and priority;
- Provide for end-to-end and link-to-link communication and network security.

The architecture envisioned a network-of-networks consisting of end users (both on Earth and in space), various relay systems, Earth-based (and possibly on other locations such as the lunar or Martian surface) infrastructure ground stations, control centers, science operations centers, and network infrastructure providers. The various elements were to be interconnected through communication links using physical connections (as is the case for terrestrial or local site in-space links), RF, and optical communications as appropriate for the environment, mission requirements, and cost/benefit analyses. This model lent itself to be considered in terms of an information flow model and a data flow model in which the information flow is overlaid on the data flow, which is itself overlaid on the various physical and operational elements that provide the communication and network management services. This model allows information exchange between users to flow logically from source to destination independent of the underlying network structure and allows the use of both direct point-to-point and multipoint/multi-hop communication links as the geography and orbital mechanics of the mission dictate.

The SCAWG based its services model on the Open Systems Interconnection (OSI) model, including the terrestrial Internet's five-layer OSI simplification (see Figure 4.3). Within this model, application services are provided to end users such as file transfer, audio and video streaming, and messaging. Application services rely on a transport layer providing application-to-application end-to-end data transfer capabilities and qualities of service such as reliability and order of arrival preservation. Transport layer services rely on their underlying network layer services that provide routing and queueing of data between applications across logical links. The link layer provides structured data transport across a single point-to-point communication link, or hop. Finally, the physical layer provides the actual communication channel, whether RF, optical, copper wire, or human courier.

Figure 4.4 depicts the overall SCAWG communications architecture decomposed into its various service layer interfaces. The end-to-end data transfer occurs between two end users (in this case one on Earth and one remote). Each end user system has a corresponding set of applications which rely on matching application, transport, and network services. Outside of the user end system the commonality becomes less important as the architecture moves into link and physical layer services. Each link layer service can be different as can the various physical layer services depending on the specific connection, medium, and system implementation that the data transfer will utilize. In this example an intermediary relay system is shown providing additional application layer services of real-time or store-and-forward

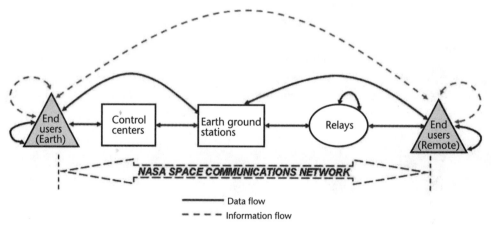

Figure 4.3 SCAWG information and data flow architecture models. (Image credit: NASA.)

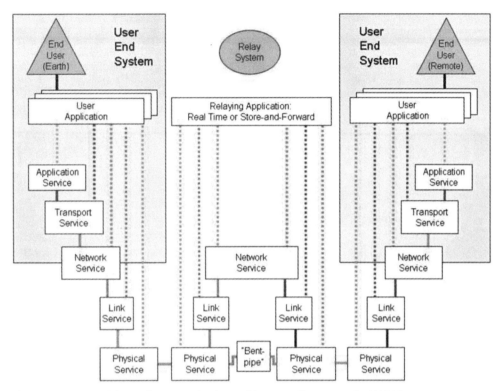

Figure 4.4 SCAWG layered communication architecture. (Image credit: NASA.)

delivery of information. Also shown is a bent pipe relay link such as those provided by NASA's TDRSS and most commercial communication relay satellites in which no higher-level application services are provided, simply serving to move information from a source to a destination.

Finally, the SCAWG recognized the need in the architecture for a method to configure, schedule, control, and manage the overall network and the individual information flows. The service management model and corresponding service management interfaces between the various end users and intermediary network nodes. Just as in terrestrial communications and networking the SCAWG's service management model needed to provide for equipment and physical systems, scheduling the use of a system, channel, or bandwidth, and establish protocols for coordinating between space communication network element providers, missions, and end users. Table 4.1 shows a simplistic comparison between the functions of the SCAWG service management model and our everyday experience with communication and data service providers such as your cell phone company or ISP.

While the SCAWG examined the narrow field of space communication (as opposed to communication and networking in general), its recommendations were taken forward by NASA to establish research and development activities in delay-/disruption-tolerant networking, high reliability transport layer protocols, ad hoc and distributed network management techniques, and to develop the operations concepts and mission requirements to drive investment in the technology and hardware needed to begin deploying the interplanetary network. The results of the

Table 4.1 Comparison of SCAWG Service Model with Real-World Services

SCAWG Model	Everyday Model
Building and deploying space communication network elements	Telephone companies and internet service providers building out cell infrastructure, customers signing service agreements when they buy cell phones, ISPs laying fiber optic cables, installing cable modems
Establishing service level agreements and commitment to provide communication services / access / bandwidth to end users and missions	Customers signing up for cell phone, cable, or internet service
Scheduling network resources (channels, bandwidth, antenna time)	Renting a satellite TV transponder, arranging for a phone call, setting how often to check e-mail
Establishing a communication session	Dialing a telephone number
Monitoring quality of service and service commitment	Tracking number of dropped calls, busy signals, failed file transfers, and dealing with customer service

SCAWG became the United States' position in NASA's dealings with its partner space agencies, and formed the core of the architecture, services, and protocol recommendations addressed by the international community in the SISG.

4.3.4 Space Internetworking Strategy Group

The IOAG SISG was established by the IOAG to explore the challenges specific to interoperable internetworking between missions, space communication and tracking networks, and ground operations systems/control centers among the various civilian space agencies. The SISG focused on finding approaches to interoperability between systems and facilities of the various member agencies that were technically realizable, operationally useful, and politically feasible. Where the IOAG membership consists of senior leadership of the various member space agencies, the SISG was staffed by technical experts appointed by the IOAG agencies in the fields of space communications, space mission operations, and space policy. The SISG was effectively the technical working group established and chartered to find solutions to the challenges identified by the senior leadership comprising the IOAG.

4.4 IOP Meets the Consultative Committee for Space Data Systems

In parallel with the efforts of the IOP and the IOAG, the CCSDS had been working to develop common, interoperable standards for space communications and tracking with the goal of standardizing how information would be described, packaged, transmitted, received, and stored by space missions of the world's space exploration community. Where the IOP (and IOAG) took a bureaucratic, management-down approach to the challenges of interoperability, the CCSDS was established and worked from a technician- and operator-up approach. Focused almost exclusively on addressing the technical and operational challenges of standardizing space communication, the CCSDS was attempting to solve problems that were not directly addressed by agency or program leadership in the form of policy and requirements. The IOP and its IOAG were tasked to address the same problems, but from a top-down, architecture, and policy focused approach. During the years in which the two organizations were developing it was found by the member agencies that combining the efforts into a complimentary structure would benefit everybody.

As the agencies were providing resources and focus for both the CCSDS and the IOP and IOAG, and as the same individuals often participated in the technical, operations, and architecture work of both the CCSDS and the IOAG, a decision was made to better align the work overall. While the CCSDS would remain functionally independent of the IOP and IOAG, the member agencies decided that the CCSDS would interact with the IOAG and accept guidance from directives resulting out of the IOP. This provided additional working expertise to support the policy development of the IOP and IOAG while at the same time providing support and credibility to the work of the CCSDS.

Following the IOP-2 in 2008, the IOP secretariat issued a communique in October 2009 to the secretariat of the CCSDS requesting that the CCSDS work toward the development, testing, and standardization of protocols for interoperability based on the architecture recommendations developed by the SISG [14]. This decision and agreement resulted in committing the member space agencies to the path of developing the necessary standards and protocols (and implementations thereof) to implement the delay and disruption tolerant network-of-networks envisioned by the SISG, resulting in much of the DTN research and technology of today.

4.5 DTN in the IRTF

The DTNRG was formed in 2002 under the IRTF "as a result of the observation that a noninteractive, asynchronous form of messaging service, able to operate over diverse types of networks, would be useful for several networks currently in use or being contemplated [15]." Work within the DTRNG was based on previous activities by the IRTF's Interplanetary Internet Research Group (IPNRG) with the exception that where the IPNRG focus was on space systems and the problems posed by deep space communications, the DTNRG expanded the breadth of discussion of DTNs to include the traditional terrestrial, mobile, and airborne environments. This resulted in a shift away from the nuances of delay/disruption tolerant networks that must operate within the physics of space flight (very long distances, intermittent visibility, orbital mechanics, high velocities, etc.) and allowed considerations of DTNs in far less challenging network topographies and dynamics. The DTNRG concluded its work under the IRTF in April 2016 and has been stood down [16].

Members of the DTNRG contributed to the development and standardization of the Delay Tolerant Networking Architecture (RFC 4838) and the Bundle Protocol Specification (RFC 5050). These protocols were used to develop the DTN2 implementation which was targeted for terrestrial DTN applications. DTN2 was also used as one of the two independent DTN implementations (along with ION) used to verify interoperability of the protocol specifications within the CCSDS.

4.6 Ongoing Development

DTN protocols and implementations for International Space Science missions and Human Space Flight use contiued development with funding and sponsorship in the United States by NASA's Advanced Exploration Projects (AES) through 2019. NASA's AES Program stood down in fiscal year 2019 with the responsibility for continuing DTN research within NASA devolving to the various programs and projects that will be responsible for implementing the technology in operational missions and systems. NASA continues to work in collaboration (and separate funding) with CCSDS with international participation in the standardization process. Software development is being conducted by a number of international space agencies, all adhering to the RFC5050 bundle specification which is called for by CCSDS.

As of this writing, in a separate effort (which is unrelated to current international space use development), a new DTN bundle protocol specification (BP Version 7) is being developed in the IETF which will not be interoperable with the current international space DTN implementations built upon BP Version 6 and baselined for the Lunar and Mars exploration programs in NASA/ESA/KARI and JAXA. Determining the upgrade path from BPv6 to BPv7, how to accomplish interoperabilty and coexistence of these protocol versions, and the future path of DTN research in general, are currently open questions.

4.7 Summary

The research into networking capabilities in increasingly challenged environments that led to the development of DTN was initiated by international space agencies as a way of increasing interoperability between and among space assets. This work aligned with similar work being performed by defense organizations for next generation internet capabilities.

International space agencies coalesced their research requirements to the CCSDS through direction from the IOAG. Under the auspices of specifying the requirements and mechanisms of a Solar System Internet, the CCSDS was given direction to develop capabilities to achieve a variety of network features in the challenging conditions of space.

DARPA coalesced their research requirements through academia and ultimately the DTN research group in the IRTF. The IRTF produced several seminal documents associated with DTN, including its architecture, RFC4838, and the BP (RFC5050). RFC5050 served as the base document for the CCSDS standardization of the BP.

Fundamental research into the transport layers associated with challenged networking has largely completed. The DTNRG and DARPA efforts in this area have completed. Space agencies have begun the deployment of DTN into their space networks and research related to shifted to advanced topics such as exotic routing techniques, autonomous network healing, and behavioral analysis.

4.8 Problems

4.1 Select two communities (space, defense, academia, and industry) whose technical issues helped start research into DTNs and describe how their problems overlapped. Explain why a DTN solution for one community would solve a problem for another community.

4.2 Explain why the interoperability goals sought by the IOP would be satisfied by an appropriately architected network.

4.3 Consider the requirements for an interplanetary network. Would you add any requirements to this list? Why or why not?

4.4 Explain in what ways the requirements for an interplanetary network differ from requirements for networking over the terrestrial internet.

4.5 Provide an example of applying the SCAWG services of monitoring quality of service and service commitment to a spacecraft.

4.6 Describe the relationships between the IOP, IOAG, SISG, CCSDS, and SCAWG.

4.7 Consider the live exercises demonstrating DTN that occurred in 2010 as part of DARPA efforts for next generation networking. Explain how the capabilities demonstrated in these exercises would provide benefits to civilian space agencies.

References

[1] Torgerson, J. L., "History Correction," received by Keith Scott, personal communication, June 30, 2015.

[2] "BBN Technologies to Stabilize Milcom Connectivity," *Satnews, Daily Satellite News*, August 20, 2008, www.satnews.com/story.php?number=2088415375.

[3] Burleigh, S., "Compressed Bundle Header Encoding (CBHE)," RFC 6260, May 2011, <https://www.rfc-editor.org/info/rfc6260>.

[4] Burleigh, S., "Contact Graph Routing," January 2011, https://tools.ietf.org/html/draft-burleigh-dtnrg-cgr-01.

[5] Burleigh, S., "Bundle Protocol Extended Class of Service (ECOS)," January 2014, https://tools.ietf.org/html/draft-burleigh-dtnrg-cgr-01.

[6] Fall, K., et. al., "The DTN URI Scheme," September 2009, https://tools.ietf.org/html/draft-irtf-dtnrg-dtn-uri-scheme-00.

[7] Torgerson, J. L., "DARPA Phase 3 DTN Core Engineering Support, Final Report," *JPL Publication Number D-65838*, Jet Propulsion Laboratory, Pasadena, CA, April 30, 2010.

[8] Interoperability Plenary Objectives, https://www.interoperabilityplenary.org/home.aspx

[9] "Minutes of the Interoperability Plenary Meeting," *Interoperability Plenary*, Paris, France, June 1999.

[10] "Communique of the Meeting of IOP-2 at the International Hotel in Geneva, Switzerland," *Interoperability Plenary*, Geneva, Switzerland, December 2008.

[11] "Communique of the Meeting of IOP-3 at the Cité de l'Espace, Toulouse, France," *Interoperability Plenary*, Toulouse, France, June 2013.

[12]   Interagency Operations Advisory Group, https://www.ioag.org/default.aspx
[13]   "NASA Space Communication and Navigation Architecture Recommendations for 2005-2030 Final Report," *SCAWG*, NASA, Washington, DC, May 15, 2006.
[14]   "IOAG-13 Liaison Statement" from the IOAG Secretariat to the CCSDS Secretariat, Interagency Operations Advisory Group, October 2009.
[15]   "About the DTNRG." Google Sites, sites.google.com/site/dtnresgroup/home/about, https://irtf.org/concluded/dtnrg, historical archive.
[16]   Delay-Tolerant Networking Research Group (DTNRG), Internet Research Task Force, April 5, 2016, irtf.org/concluded/dtnrg.

CHAPTER 5

Where the Terrestrial Internet Is Not Enough: Motivating Use Cases

The terrestrial internet accommodates so many use cases that its capabilities have become synonymous with the requirements of computer networking. While these capabilities continue to expand and evolve, there are several motivating use cases that are not well accommodated by this terrestrial paradigm. This chapter presents motivating use cases for DTN architectures and technologies that representing implementations deployed in real-world (and off-world) applications.

5.1 The Value of Use Cases

The initial motivator for research into DTNs were grounded in a desire to recreate a terrestrial internet experience in environments that are unable to deploy the terrestrial internet infrastructure. This desire to pervade existing networking operational concepts stems from the perception that minimizing changes to a proven system reduces system cost and risk. However, if the fundamental characteristics of a new system are incompatible with an existing system, minimizing changes in the new system can increase the cost and risk of the system as incompatibilities are discovered, and patched, as part of system deployment and early operations.

The acts of network architecture design, application development, and operations need to be mapped to the unique characteristics of the system being deployed. To help understand these characteristics, this chapter presents a series of unique use cases in a variety of domains. The goal of these use cases it to help differentiate challenged networking conditions from terrestrial internet conditions and, in doing so, appreciate the need for designing new technologies, architecture, and software implementations.

5.2 The Solar System Internet

The space missions of today rely heavily on networked communications at the surface of in-space destinations, across vast distances of space, and terrestrially among control centers, ground stations, science centers, and the public. Rovers on Mars communicate with Earth via orbiting relays, Earth orbiting spacecraft talk directly to ground stations, each other, and relays in geosynchronous orbit, and deep space missions communicate with Earth from beyond the edge of the solar system.

5.2.1 A Brief History of Space Communication

Space communications began as a strictly point-to-point, line-of-sight capability in which a single spacecraft would communicate with a ground station when both were in view. If the ground network was sufficiently sized, and the mission sufficiently prioritized, the spacecraft would take another pass as one ground station moved out of sight and another came into view. The model for traditional space communication is very much that of a telephone operator at a switchboard making and breaking connections on a preplanned schedule. If both parties are there when the connection is made a conversation can happen. If one party isn't there or the connection isn't made at the right time the conversation fails.

Figure 5.1(a) is a picture of NASA's Deep Space Network 34m antennas in Canberra, Australia. Each antenna is very capable and can transfer many hundreds of megabits per second to and from Earth orbit and hundreds of kilobits to megabits between Earth and Mars. As spacecraft move deeper into the solar system, the achievable data rate drops off significantly as a result of the RF energy spreading geometrically as the distance increases. Additionally, the antennas have very tight beams and can only be pointed in one direction. Most (high) Earth orbiting space craft are only in their field of view for a matter of minutes. (Lower orbiting spacecraft cannot be tracked by the DSN antennas and must use smaller, faster near-Earth antennas instead such as the 11m antenna at NASA's Wallops Flight Facility in Virginia shown in Figure 5.1(b)).

As spacecraft move further into the solar system, the periods of visibility become longer, but each antenna complex can only stare at a given location for approximately 8 hours, due to the rotation of the Earth. Given the orbits and trajectories of the satellites, the motion of the Earth, and the sparse placement of the expensive and maintenance-intensive antennas this provides any given near-Earth space mission with only a few passes of a few minutes per day (with high-speed tracking antennas), and typically a single pass every few days for deep space missions. This is hardly the near continuous communication we have come to expect in our daily lives. The availability can be improved by building more antennas or dedicating them to specific missions, but the cost and operations complexity increases rapidly.

This lack of reliable connectivity means that spacecraft are often designed with large recorders to store mission and science data because they are not always in view of a ground station and even if a pass is scheduled it is possible for a failure or conflicting priorities to cause the pass to be missed. Operationally, these handoffs must be both planned and managed in real time, driving the work required to support a mission. The managing of these events takes place on one part by the service provider and a second part by the mission planner or flight control team. Even with computers and modern databases this is a time and effort intensive process.

The introduction of dedicated networks of geosynchronous relay satellites for commercial and space exploration applications improved this model by allowing spacecraft flying below the geostationary equatorial orbit (GEO) belt to

5.2 The Solar System Internet

(a)

(b)

Figure 5.1 (a) NASA's 34m DSN Ground Stations at Canberra, Australia, and (b) NASA's 11m Ground Station at NASA's Wallops Flight Facility, Wallops Island, VA. (Image credit: NASA.)

communicate with the Earth via relay through dedicated networks of communication satellites in orbit above them. For example, NASA's Tracking and Data Relay Satellite (TDRS) [1], along with dedicated ground sites and ground network connectivity, make up NASA's Space Network (see Figure 5.2).

In this networking model, a user mission communicates through another satellite located in geosynchronous orbit which relays the communication to a ground station for distribution to the mission's control center. Figure 5.3 provides a schematic representation of NASA's Space Network providing communication between a user mission (called a customer platform by NASA) and the mission control center (the customer ground facilities) [3]. Since the relay satellite is continuously in

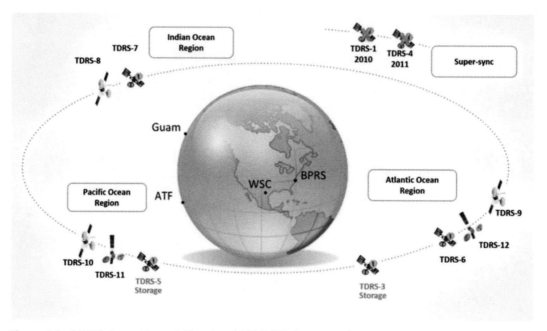

Figure 5.2 NASA's Space Network Fleet (as of 2015) [2]. (Image credit: NASA.)

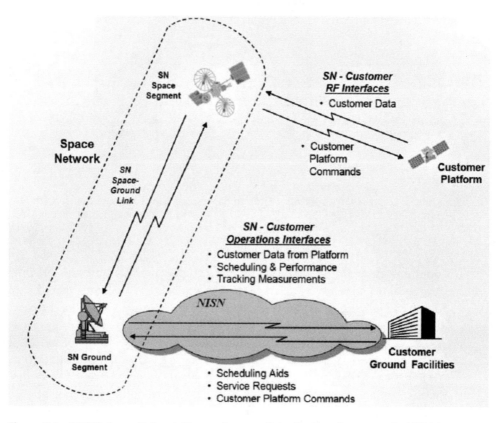

Figure 5.3 NASA's Space Network Geosynchronous Relay System. (Image credit: NASA.)

view of its dedicated ground site and in a very high orbit, it can see a much larger region of space in the orbits below. This provides high availability for point-to-point relay links between the user mission spacecraft and its control center.

With the launch and deployment of GEO-based communication satellites and networks like TDRS and the Space Network, most of Earth's orbit could be geometrically accessed at any time through relay in addition to the higher performance but more geometrically constrained and less available ground stations. While TDRS is among the most capable geosynchronous relay satellites developed, it and other relay satellites like it remain limited in the number of missions that can be simultaneously supported due to only having a set number of main customer antennas and/or limited size multiple access phased array. Figure 5.4 depicts the third-generation design of NASA's TDRS spacecraft showing the two main customer antennas and the body mounted phased array (referred to by NASA as the single access antennas and the multiple access antenna respectively.) Relay satellites can be designed with a great number of customer antennas but are fundamentally limited due to size, weight, and power restrictions. Even with an infinite number of antennas, each antenna still must be scheduled, pointed, and track the moving customer spacecraft to maintain the communication contact throughout the pass.

This significantly improved the connectivity and availability of communication, but still did not provide real-time continuous communication to all but the most critical missions (not even the now retired Space Shuttle or the International Space Station have continuous communication with Earth). Additionally, the operations model remained that of a switchboard operator with a schedule as each pass had to be planned and the communication asset—whether ground station or relay—scheduled, pointed, configured, and connected to the terrestrial data path.

In 2003, NASA's JPL began experimenting with the idea of store-and-forward routing and relay delivery of data between Earth and robots on Mars. Around

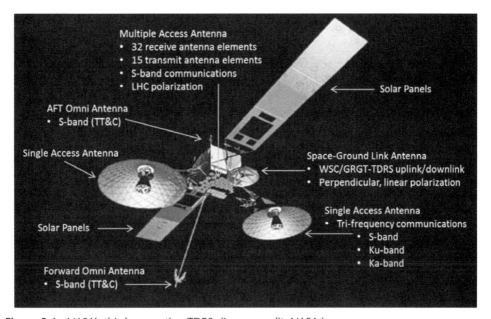

Figure 5.4 NASA's third-generation TDRS. (Image credit: NASA.)

the same time, NASA's Goddard Space Flight Center began to examine the use of terrestrial internet protocols and possible adaptations to enable networked communication in near-Earth and the lunar environment. In neither case were these traditional protocols well suited for the task. Furthermore, effective protocols were not well-defined, and a network service model did not exist.

Just as the traditional telephone line moved from manual operator switchboards and analog voice communication through the era of automatic telephone switching and into today's fully networked communications (where voice calls are actually sent as digital data across IP routed networks) a similar evolution in thinking about space communication took place. As space missions continued to increase in complexity, mission planners and space communication service providers (typically the civilian space agencies) looked to the advantages of advancing network technologies with a desire to replicate if not actually duplicate. Space communication providers began to consider themselves more as network providers along the lines of commercial ISPs rather than link providers like the old school telephone companies. These space ISPs and mission designers took the obvious step of looking to the terrestrial internet as a model for highly effective, efficient, reliable, and adaptable data communication. What they quickly found was that taking the internet from Earth's surface into space introduces several new challenges and forces the network to operate in an environment that is very different from that found terrestrially.

During the late 1990s and early 2000s, the number of orbiting and landed missions at Mars began to increase. Both NASA and ESA successfully landed rovers on the surface and several orbiting Mars observing satellites were in place. While not originally intended for this purpose, a team at JPL began experimenting with repurposing the radios of the surface and orbiting vehicles to communicate with each other. Making several software hacks, the engineers were able to transfer files between lander and orbiter and then on from orbiter to Earth. The links were manually configured and moving the data was done entirely through scripting each event, but the data did move. This early experimenting came to be known as the Mars Network.

5.2.2 A Solar System Internet

Following the experiments with the Mars Network, the idea for a Solar System Internet (SSI) was first proposed in a memo in the InterPlanetary Networking Special Interest Group (IPNSIG) [4] and shortly thereafter described to a wider audience in a 2003 paper by internet pioneer Vint Cerf, JPL engineers Scott Burleigh, Adrian Hooke, and Leigh Torgerson, Kevin Fall of Intel, Bob Durst, and Keith Scott of the MITRE Corporation, and Howard Weiss of SPARTA, Inc. [5]. The paper began by discussing the challenges of deep space communication and robotic mission teleoperation and why traditional terrestrial internet protocols of the TCP/IP suite perform poorly in such environments. In the final analysis, TCP/IP fails because of a few fundamental assumptions that are not factors in traditional networking. TCP/IP assumes richly connected, low latency, simultaneously bidirectional links.

5.2.2.1 Connectivity

Networking in terrestrial (and most airborne) communication often assumes and depends on nearly continuous connectivity. The protocols that make up the internet TCP/IP suite don't just assume that nearest neighbors are able to communicate, but that an end-to-end connectivity spanning multiple hops exists as well. Infrastructure routers are typically connected by physical copper or optical fiber cabling or via dedicated RF/microwave paths. A great deal of investment and effort by the telecommunications industry goes into the planning, construction, and maintenance of these physical connections. Companies and individuals wire their offices and homes and pay for always on connections to wide area communication channels in the expectation that when they want to move information the path will just work. Edge routers connect to end users either through dedicated cabling (such as the Ethernet cable in an office network) or highly available RF wireless links (like the Wi-Fi signal in your home or hotel room). Where networking nodes are not available, most users can leverage the commercial cellular network to send IP data. An end user can therefore almost always assume that wherever they need network connectivity it will be available in some form. To the industry's credit, this has become such a reality that it is a frustrating annoyance when this expectation is not met and our cell phone doesn't have enough bars for a good signal!

The situation is very different in space. Geometries and orbital mechanics drive whether or not a spacecraft is in view of a communications relay or ground station at any given time or whether there is a planet, moon, or other body in the path. Where geometries are not the limiting case, resource availability and utilization often drive the availability of a given link. Unlike the nearly ubiquitous commercial telecom industry providing landline, microwave, and cellular telephone and data infrastructure, space communication infrastructure is very specialized and not common. A space agency may only have a handful of near- or deep-space communication assets and while a commercial industry exists to provide space communication it is not large. So, resources are scarce to begin with.

Making the availability issue more challenging is that unlike a cellular tower or an ISP's network connection, if a ground station or in-space relay is pointed at a spacecraft at Mars it cannot also be pointed to one at Venus. The two contacts must occur at different times. (While some space communication assets can communicate with more than one user at a time this is not common and even in these cases the systems are typically limited to only supporting a handful of simultaneous users that must often be in similar locations.) These space communication infrastructure elements must therefore be carefully allocated and scheduled to support communication to missions with each pass consisting of changing the physical direction in which the asset is pointed, how it is configured, what frequency, modulation, and coding will be used, and so forth. This has two implications: first, there will likely be a delay between when a spacecraft collects or generates data which it needs to transmit and when the physical communication link is available; and second, it is very possible that there is rarely—or never!—a continuous path between any two nodes in space. Without a continuous, end-to-end path available, traditional TCP/IP protocols fail to operate.

5.2.2.2 Latency

Where terrestrial networks typically have low latencies (of the order of tens of ms) between routers and those due more to the processing speeds of the hardware than the physical distance between them, space communication links extend into the billions of miles (Pluto is 3.67 billion miles from the Sun). Since RF and optical communication signals travel at the speed of light, these distances cause large delays in data transmission from source to destination (more than five hours from Earth to Pluto). These ranges also change with the motion of the planets, asteroids, and other bodies in space causing a wide variation in delay from closest approach to farthest distance. For example, at its closest approach to Earth (54.6 million km), the light speed round trip time (RTT) for the Earth-to-Mars path is about 7 minutes. 3.5 minutes Earth to Mars, and 3.5 minutes for the return trip from Mars to Earth. When Mars and Earth are on opposite sides of the Sun (about 401 million km) that RTT increases to around 40 minutes. At the same time, surface explorers and orbiting constellations of spacecraft may be in very close proximity (meters to thousands of km) making the range of time delays experienced across the solar system a span of 12 orders of magnitude (10^{-8} to 10^4). While standard protocols can be tuned to account for either very long or very short delays, tuning for 12 orders of magnitude (a factor of 1,000,000,000,000) is impractical. IPs were not designed to efficiently handle this wide variation of delay, and at a certain point as delay increases traditional terrestrial IPs will fail entirely.

Making the situation worse, for protocols that assume short latencies (such as TCP or SMTP) this becomes a real problem. If a protocol assumes that it should receive a response (positive or negative) within a certain period of time, and that time is exceeded, the protocol will assume the transmission has failed and will attempt some form of correction usually involving a retransmission of the data and/or a throttling of the data rate. As the latency of the link increases a protocol designed this way becomes less efficient until eventually it fails completely assuming (incorrectly) that the data never made it and neither did the request for confirmation. In reality, the data is still in flight through interplanetary space somewhere between Earth and Mars, and any possible response from the other end will come back outside the window of tolerance for the protocol. The implication of this is that whatever protocol is developed to operate in a high latency (delayed) environment must not assume that a response will be received within any set time—and that a response may never be received.

In addition to free-space latency due to the speed at which RF and optical signals propagate, the links between two nodes in space may not always be available. If the link will eventually be available, this delay due to the lack of a path can be considered a connectivity disruption—albeit a very long one. In a certain sense any disrupted link can be thought of as a network delay where the delay corresponds to the time required to obtain a ground station or relay's service and to come into visibility and range to establish the link. If that link will never again be available for some reason, the delay effectively becomes infinite!

5.2.2.3 Bidirectionality

Where disconnected links and long latencies pose challenging but obvious problems, an assumption of link bidirectionality introduces problems that are both subtle and often more challenging to solve. Traditional networking (and communications in general) generally assumes a simultaneously bidirectional flow of information. When we have a conversation, we expect to hear something back from the person we are speaking with and we expect to hear from them almost immediately after we have said something (if they don't just talk over us at the same time!) TCP/IP based networks (as well as those based on other chatty protocols) assume the same thing. The TCP/IP protocol suite assumes that either end of the link can send information in either direction and can do so at the same time and generally the same rate (to within a few orders of magnitude). The TCP protocol uses this assumption to provide reliable transfer (by acknowledging receipt of an IP packet or requesting a retransmission in the event of a failure) and channel rate throttling to reduce errors in the link when they begin to occur above some threshold frequency.

TCP is a chatty protocol in that several messages are sent in the reverse direction for each transfer of useful data in the forward path. The TCP/IP suite also assumes (nearly) balanced bandwidths in the forward and reverse direction implying that the protocol does not need to be concerned with how much traffic it generates in the reverse direction as usually more information is being sent forward (the IP packets containing the data) than is being sent back (the control and protocol messages.)

5.2.2.4 Asymmetry

The assumption for bidirectional communications fails in the space environment and so do traditional networking protocols. Why exactly is this?

First, unlike most terrestrial network links, in space communications a forward path is not always paired with a return path. Space communication employs the concept of uplink, which generally refers to the transmission of information from Earth to space, and downlink, which is in the space to Earth direction. Links between two systems in space are generally referred to as crosslinks and the directions are referred to as forward and return relative to the source of the information. (Both spacecraft can have forward and return links between each other at the same time. They are the same physical link just reversed in direction depending on the perspective of which is the source and which is the destination.)

Space links are planned independently and while there are often simultaneous forward and return links (or up and downlinks), there is no pairwise association between the two. They are actually two totally independent links that happen to occur at the same time. The links can and often are actually provided by different ground stations for uplink and downlink further decoupling the two communication channels. Spacecraft often operate with only an uplink (say to receive commands from their control center) or only a downlink (to return collected science or mission information). Often missions will have several downlink sessions to offload data from storage before the next uplink session as the purpose of most missions is to collect science information. There is usually very little to send to the spacecraft and so it is not efficient use of the limited antennas to always have an uplink for every downlink.

Second, the two links are often operating at very different data rates. Uplinks typically send much less information and are generally more difficult to achieve as the signal from Earth must be received on a relatively small, low-sensitivity spacecraft antenna while a downlink signal received from a spacecraft can be collected with large aperture, very sensitive antennas such as those of the Deep Space Network. Also, the main function of the mission is to bring back information and so much more effort is placed on designing higher bandwidth downlink/return links than uplink/forward links. It is simply not a good use of resources to design for high rate uplinks at the expense of the ability to send back large volumes of science. These two factors—physics and the purpose of the mission—mean that the downlink is very often many orders of magnitude greater in bandwidth than the uplink. The implication for a chatty protocol like TCP is that the available bandwidth for protocol management traffic can be quite limited and often cannot accommodate the required volume as the rate of downlink data increases.

While this discussion has addressed space communications primarily, similar situations often occur in terrestrial and airborne scenarios. From a networking perspective the challenges to be overcome are the same and so lessons learned from examining space communication use cases can often be applied to other resource-constrained architectures and environments.

5.2.3 An Architecture for the SSI

Lack of reliable connectivity, presence of large and variable latency, and lack of simultaneous and balanced forward and return links spell doom for traditional network protocols that were simply not designed for this environment. This was recognized by Cerf et. al., and after addressing the shortcomings of traditional TCP/IP in the space environment, the paper proceeded to describe the needed behaviors and characteristics of an envisioned new protocol suite that would be suitable for a high delay, low bandwidth, highly disconnected communications environment, and proposed an architecture based on bundling of data across an overlay network providing end-to-end delivery of data in a challenged environment. Figure 5.5 depicts a DTN network overlaid on a generic model of heterogenous transport, network, link, and physical layers across deep space.

Such a network would provide the best of traditional communications including packetizing information and routing it based on source and destination, transferring the burden of moving the information packets from the end users (spacecraft, control centers, etc.) onto the network itself, and the ability to reliably move information across multihop topologies with assured qualities of service while

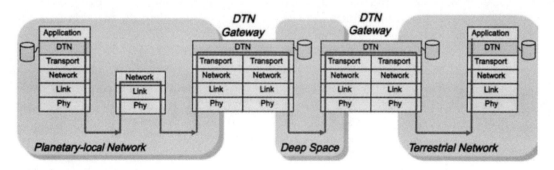

Figure 5.5 DTN as an overlay network.

taking advantage of the variety of communications links available (physically hard connected, one-way RF links, IP routed paths, space-customized protocols, etc.) The key concept was to build a network that could be overlaid onto whatever underlying infrastructure, protocols, and communications were available, using what worked well on a link-by-link and hop-by-hop basis while providing an end-to-end capability that is independent of the assumptions that cause traditional networking to fail in space. This was the approach taken by engineers in developing the Solar System Internet concept.

As was soon realized, the concept of bundling data and providing specialized protocols to handle the moving and accounting for bundles across a heterogenous network of IP networks, deep space links, and local radio communications had merit and the group began to investigate the problem in detail. The space community—particularly NASA's JPL and Goddard Space Flight Center (GSFC) facilities—formed research teams to work out the needed architecture, operations concepts, technologies, and protocols that would be required to make such a bundle overlay network a possibility. While there was disagreement between the various research teams, this was confined primarily to technical debates on specific protocols, network behaviors, and data formatting. The concept of the interplanetary internet had universal appeal.

As scientists, engineers, and mission planners began to address the challenges of networking in a space environment the concept of a network of networks or a network of interconnected internets emerged. This model treats a solar-system-wide network as if it were islands of various levels of connectivity and delay with the gulfs of deep space spanned by high-performance point-to-point links. The architecture very much began to resemble a hub-and-spoke or star topology. This network model is often used in connecting remote offices with a corporate headquarters or various buildings on a college campus. In this case it is only the physical scale of the network that is different—and along with that scale comes unique challenges.

Figure 5.6 depicts the concept that emerged within the NASA teams and the IPNSIG of regions of network connectivity each tuned for their specific performance environment and the basic physics of their orbits, geometries, and links. The various regions would then be interconnected by an infrastructure of internetworking that would provide the bridge between the various islands of networked communication. As the architecture evolved it became clear that an overlay network such as that envisioned by Cerf and the early DARPA Interplanetary Internet team would suit the network interconnection well so long as a set of protocols could be developed that would not fail in the long-haul, deep-space communications environment where traditional IPs did. The main factors in which traditional internet protocols fail in the space environment are results of simple physics and manifest in challenges in latency and connectivity: delay and disruption. What was needed was a DTN approach.

5.3 Distributed Spacecraft Constellations

The idea of constellations of spacecraft working in concert to accomplish a mission or provide a service has developed over the past 20 years from drawing boards and

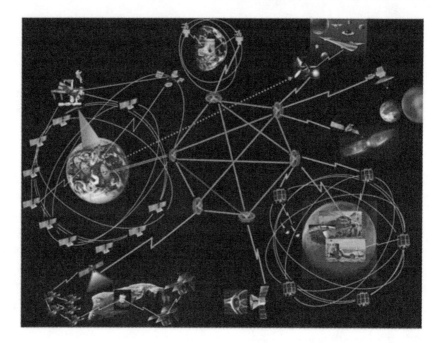

Figure 5.6 Artist depiction of the Solar System Internet consisting of a network of internets. (Image credit: NASA.)

white papers to realized implementations. Constellations of spacecraft offer several advantages over single spacecraft—and some applications can only be met by multiple spacecraft flying simultaneously. While mission have been (and are being) flown with limited inter-spacecraft and intra-constellation connectivity, great advances in capability and functionality can be realized if the individual space vehicles can be provided with reliable internetworking within and among the constellation members, control centers, and end users.

5.3.1 Planetary Observation Missions and Space-Ground Integration

In the science and space exploration domain, constellations of spacecraft have been envisioned for Earth and planetary science observation, communication and navigation support to lunar, asteroid, planetary, and deep space missions, and novel instruments based on precise formation flight allowing mission teams to create instruments and sensors that would not be physically or practically realizable as a monolithic vehicle.

Observation of planetary bodies—to include NASA's mission to Earth—are often made up of multiple independent spacecraft performing remote sensing of the planet they orbit. These satellites carry instruments such as radars and radiometers to detect structures and electromagnetic phenomena and the energy balance of the planet. Imagers are flown in the visible, infrared, and hyperspectral bands to detect, measure, and characterize the biome and water cycle of the planet. Instruments such as laser interrogators and lidars are used to obtain precise measurement of surface features, characteristics of ice (water or another chemical), and measure turbulence in the planetary atmosphere. In order to truly study a planet,

5.3 Distributed Spacecraft Constellations

it is typically necessary to both overfly the entire surface to get a global picture and to repeat the same measurement in the same location many times to learn how changes occur over time. This is true whether the mission is to measure the climate change of Earth, to monitor the weather of Mars, or map the changing geography of active worlds deep in the solar system.

A lesson learned through years of operating NASA's Earth Observing System (EOS) is that coordination between spacecraft with different capabilities provides a robust opportunity to capture events that would otherwise be missed. NASA's A-Train and C-Train of Earth monitoring satellites is a fleet of Earth observing satellites that fly in a follow-the-leader orbit, such that each spacecraft passes over the same location on Earth one after the other in close proximity. The afternoon orbit, or A-Train, consists of NASA's Orbiting Carbon Observatory 2 (OCO-2), JAXA's Global Change Observation Mission Water (GCOM-W), NASA's Aqua spacecraft observing the Earth's water cycle, and NASA's Aura spacecraft, which looks at Earth's air quality and climate (see Figure 5.7). Combined, the A-Train provides a global look at the Earth's climate. The C-Train includes CALIPSO and CloudSat. Each satellite carries a different instrument suite, but because of the synchronized orbits it is possible to correlate observations closely in both time and location.

A challenge for operating constellations of observation satellites is how to decide when a detected event is interesting enough to supersede existing planned observations and redirect the instruments off track, and if so, how to communicate that new tasking to coordinate the observation in a timely manner. The ability to communicate between spacecraft without direct communication links offers the ability to detect and characterize interesting events, make prioritization decisions, and provide near-real time tasking and coordination to the follow-on spacecraft so that a coordinated measurement or observation can be made on the fly.

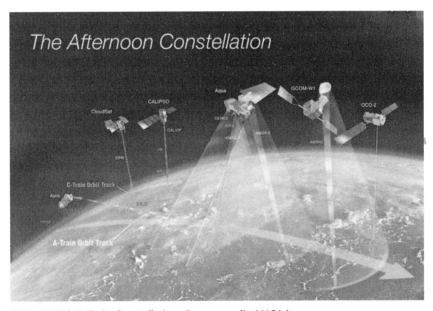

Figure 5.7 NASA's A-Train Constellation. (Image credit: NASA.)

5.3.2 Deep-Space Instruments

The ability to communicate and coordinate between spacecraft without involving Earth or manual control centers and coordination allows missions that could not otherwise be developed. Two key areas of research in mission concepts are formation flight and distributed instruments. In formation flight missions multiple similar or identical spacecraft are flown in precise geometries relative to each other. For such mission the position and attitude of the individual spacecraft must be coordinated tightly to perform the mission. Two examples of this concept are the Micro-Arcsecond X-Ray Imaging Mission (MAXIM) and Terrestrial Planet Finder (TPF) missions.

5.3.2.1 Distributed Instruments: MAXIM

Several National Academy survey reports identified the study of the origin, composition, and structure of black holes as one of the national priorities for space science decades. These objects are conducive to study in the X-ray band. Unfortunately, it requires incredibly sensitive and precise instruments to image black holes with enough resolution to explore their small-scale structure. Measurements on the order of microarcseconds are required to obtain the desired resolution.

To accomplish this, rough calculations indicate that an aperture of the order of 200m will be required. It is simply not possible to design and build a monolithic spacecraft with a large enough collecting/focusing aperture to achieve this. Nor is it possible to build a similar instrument on Earth as X-rays do not penetrate the atmosphere well and if they did it would be near impossible to stabilize the aperture and back-end diffraction, interferometer, and detection systems well enough to make the required measurement.

The purpose of the MAXIM concept (see Figure 5.8) is to make incredibly precise (micro-arcseconds in effective beamwidth) measurement of distant spade objects in the X-ray band using a distributed constellation of free-flying spacecraft in precise formation. By precision flight it is possible to synthesize an extremely large virtual structure allowing the constellation of vehicles to combine to form a single instrument that would be impossible to construct as a single object (see Figure 5.8). The overall MAXIM instrument results in a very large (15 km long) X-ray interferometer constructed out of a constellation of four different classes of spacecraft. The overall instrument consists of a focusing ring of spacecraft acting as distributed X-ray diffractors that focus incoming X-rays through a converger spacecraft, in which the wavefronts are interferometrically combined and focused onto a distant detector spacecraft. A central hub spacecraft provides a navigation and pointing reference as well as coordination among the fleet of formation flyers.

In addition to the difficulties of the science payload(s), the constellation must be coordinated such that each element's navigation and control system is precisely synchronized across the constellation. This requires the ability to communicate both pointing and attitude direction to each element (with each vector being different depending on where the specific spacecraft is in the constellation) and to do this in true real time.

Traditional point-to-point space communication techniques are simply not capable of supporting this mission concept. While an Earth-based ground station can

5.3 Distributed Spacecraft Constellations 83

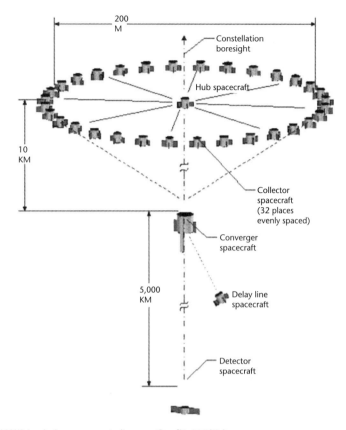

Figure 5.8 MAXIM mission concept. (Image Credit: NASA.)

(and likely would) be used to provide general tasking of observations, health and status monitoring of the overall instrument, and receive collected science data, it is not possible or practical to manage the number of individual spacecraft required. Additionally, the precision required to coordinate the navigation and attitude of the various elements cannot be achieved through links with latencies that would be experienced by the MAXIM instrument (likely deployed at a Lagrange point) attempting to communicate with Earth. This implies the need for a reliable, adaptable communication architecture local to the instrument itself that can provide interconnected communication between the in-space elements while also allowing for the deep-space communication links for command/control and science telemetry.

A network model that considers the entire MAXIM instrument as a local network that is then tied into a broad network-of-networks containing MAXIM and its Earth-based ground stations, mission operations, and science operations centers enables this concept.

5.3.2.2 Formation Flight

The TPF is a conceptual mission to detect, image, and characterize Earth-like (terrestrial) planets throughout the galaxy. The TPF was a mission concept previously under study by NASA. Consisting of a coronagraph to image and characterize the spectrum of distant start and interferometer to detect and image planetary bodies in those star's systems TPF would study extra-solar planets from their formation and development in disks of dust and gas around newly forming stars, to the presence and features of those planets orbiting the nearest stars. Ideally TPF would answer the question of whether there are potentially life-supporting planets in our stellar neighborhood.

Three TPF designs are proposed. TFP-I/Darwin (see Figure 5.9) consists of a large interferometric telescope consisting of multiple small focusing reflectors focusing their collected light on an aligned collector spacecraft. Using interferometric nulling TPF-I/Darwin would be able to remove the brightness of the star at the center of the observed solar system allowing the detection and imaging of any planets orbiting

Similar to MAXIM, TPF-I/Darwin relies on the ability to coordinate the navigation and pointing of several free flying spacecraft with very tight tolerances. TPF-C (see Figure 5.10) consists of a large monolithic telescope with a sensitive coronagraph to directly image distant stars and their planetary discs. TPF-O consists of a telescope spacecraft flying in formation with a sun shield, which is used to diffract the light from distant stars and ideally observe the occultation of those stars by planets crossing the stellar disk. TPF-C and TPF-I depend on the orientation of the angle of the distant solar system's ecliptic plane in the hopes of catching an extrasolar planet crossing the solar disc, while TPF-I/Darwin would allow true imaging of the region of space surrounding the remote star regardless of the orientation of the ecliptic. This is a good example where formation flying (enabled by local in-space networking) provides significant advantages over more monolithic spacecraft that do not depend on inter-spacecraft communications.

5.3.3 In-Space Communication and Navigation

The in-space communication and navigation use case is perhaps the most natural application of DTNs and the interplanetary internet. Just as relay satellite and

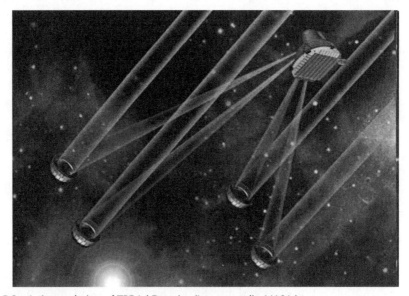

Figure 5.9 Artist rendering of TPF-I / Darwin. (Image credit: NASA.)

5.3 Distributed Spacecraft Constellations

Figure 5.10 Artist rendering of TPF-O concept. (Image credit: NASA.)

orbiting global navigation satellite systems (GNSSs) provide for global communication access, position determination, and time distribution on Earth and in Earth orbit, the same capabilities will be required as human and robotic exploration proceeds into the solar system. Architectures to provide local communication and navigation have been explored for the Moon, Mars, and other destinations. Such architectures typically resemble Earth-based relay and GNSS in that they consist of a constellation of dedicated comm/nav spacecraft in orbit of the served planetary body. Where the architectures differ significantly from Earth-based systems is that it is impractical to build a corresponding ground network of control centers and ground stations on the Moon or Mars.

As a result, concepts for in-space communication and navigation systems must rely on communication links between the C/N assets to provide services between users that are not in line-of-sight as well as to link the in-space C/N network with Earth and other destinations. NASA's Space Communications Architecture Working Group identified the following six behaviors common to all the in-space communication and navigation architectures studied (as documented in the SCAWG Final Report):

1. Communications links to/from user spacecraft;
2. Onboard routing services between user spacecraft across the in-situ C/N system as well as between user spacecraft and Earth;
3. Store-and-forward capability to provide for delayed delivery of information between user spacecraft that are not currently in line of sight or when next hop delivery is not available in real time;
4. Time synchronization and distribution within the in-situ network and with user spacecraft;
5. Time synchronization between the in-situ network and Earth;

6. Architectures for in-space communication relay networks often also include the ability to provide cryptographic and other security capabilities.

Specific concepts and recommendations of the SCAWG are discussed in Chapter 4.

5.4 Distributed and Mobile Sensor Webs

The concept of a sensor web (see Figure 5.11) was proposed by a team of researchers at NASA's JPL in 1997 [6]. The concept was developed to increase the throughput and flexibility of large sets of deployed sensor nodes by establishing communication/network links between the nodes and then allowing collected data (and in the other direction commands) to flow across the network hoping from sensor node to sensor node until the data finally reached the end user. Sensor webs were envisioned for monitoring fields, wildlife surveys, climate and pollution monitoring, geological and tectonic monitoring, ocean conditions, and deep-space science.

More than a set of distributed, independent sensors, when networked the individual nodes make up one integrated system whose elements can share data among themselves and act as one single system. A sensor web would not be necessarily restricted to a defined number of nodes or to operate within a predefined area. The primary nodes could be located anywhere in the network, and sensor webs could be deployed over time with new nodes entering the web as they are added. Multiple sensor webs deployed in proximity to each other would connect and integrate into a single larger web.

NASA and its international partners continue to explore sensor web concepts by integrating ground, airborne, and space-based systems via distributed networks. Figure 5.12 depicts NASA's SensorWeb network as of 2019.

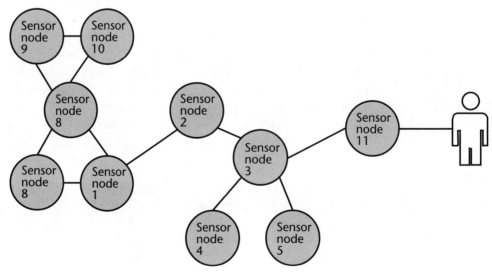

Figure 5.11 Example node and link topology of distributed sensor web.

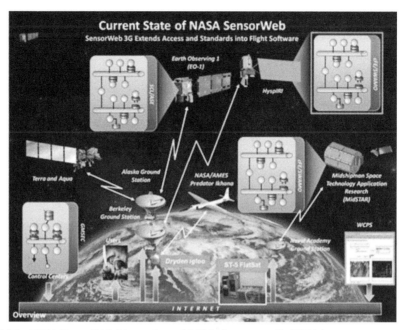

Figure 5.12 NASA SensorWeb Network as of 2019. (Image credit: NASA.)

5.5 Optical Communications

Optical (laser) communication offers significant advantages over traditional communications in that it supports much greater data rates, does not compete for already congested RF spectrum, and is very hard to intercept. Optical communication links can exist between Earth, space, aircraft, and even into deep space. NASA has operated an optical communication system from the moon and is building plans to add optical communication to the deep space architecture in the coming years. Many commercial communications providers are exploring optical communications to provide network backbones in space and governments around the world are looking to optical communications to provide the next generation of high data rate communication systems (see Figure 5.13) [7–9].

The significant disadvantage of optical communications is that lasers are blocked by clouds. To overcome this, optical communication system designers must plan for frequent but unknown periods during which weather will degrade or entirely block a laser link. This is can be addressed in several ways. First, the designers of the architecture can include many more ground sites than would normally be required on the assumption that if enough ground sites are in view there will be at least one which has an acceptable view for the laser. This helps to overcome periods of true cloud cover when a link cannot be established but it does not solve the problem of intermittent optical links during a contact as is experienced when smaller clouds pass in and out of a beam during a contact. To address the problem of intermittent links (called ratty comm), it is necessary to use a reliable link protocol and to have some form of accounting and retransmission capability when data is lost on the link. With the data rates achieved by optical communica-

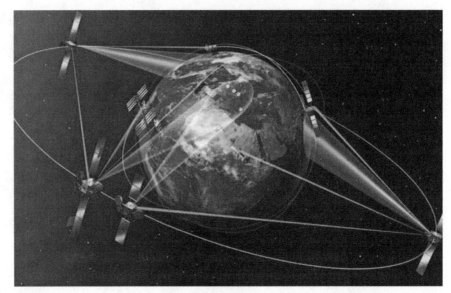

Figure 5.13 Artist concept of the European Data Relay Satellite System (EDRS). (Image credit: ESA/Airbus.)

tions (many Gbps) a significant storage and store-and-forward capability is needed onboard the relay spacecraft.

What becomes quickly apparent is that a need exists for an architecture in which the routing of information can be changed based on uncertain physical links, and that those physical links may themselves be error-prone. The network must address disruption of its connectedness and topology (see Figure 5.14), as well as provide assured end-to-end delivery. DTNs have these characteristics and are being

Information blocks flow from the satellite to the ground station and are routed to the end user. The optical link has sufficient bandwidth but the ground link is limited.

The optical link is interrupted (by a cloud). Data continues to flow from the 1st ground station through the network to the end user.

A new optical link is established with a second ground station and information is flowing again—this time with a different routing path to the end user.

Figure 5.14 Disruption and handoff of optical links caused by atmosphere.

5.6 Ad Hoc Network and Data Mules

explored by governments and industry as solutions to the challenges of free space optical communication.

Ad hoc networking is a network architecture in which there is little a priori knowledge of the members or topology of a network or of both. In these networks, members come and go with links and routes being established as they become available. When the nodes are no longer linked to the network the routes to and through them are removed. Once removed, nodes may reconnect to the network in the future, or may never see that network again.

The concept of a data mule comes from the idea of travelers using mules or other pack animals to carry not just their possessions but also messages from one location to another. Often senders in the starting location would give messages or packages (sometimes paying for the service) to travelers to deliver to recipients at the travelers' destination or at some stop along the way. The U.S. Pony Express operated this way, with riders handing off mail bags and routing between waystations and outposts until the post made it from the sender to the recipient.

In a similar fashion, information can be transferred digitally using the idea of a data mule. In this scenario the sender (or sending network) provides one node with all of the message traffic destined for a remote network. The data mule then physically or logically moves from one network to the other (see Figure 5.15), breaking its connection to the sending network region and establishing new connectivity to the second. Once established in the new network, the transported message traffic is free to continue its path to the eventual destination.

5.7 Summary

Architecture, software, and training reuse can significantly reduce the cost and risk associated with the development of new capabilities. As the desire to instrument and explore our environments (on-world and off-world) increases, new capabilities must be developed and, in such development, reuse must be considered.

Several use cases exist for which computer networking architecture, software, and training reuse are not practical. Networking across the solar system, distributed spacecraft constellations, mobile sensor networks, optical communications, and ad hoc networks all represent areas where traditional networking techniques fail. Attempting to reuse architecture, software, and training/operational concepts in these areas will likely increase the cost and schedule risks associated with these new capabilities.

5.8 Problems

5.1 Identify three differences between a terrestrial internet connection and a connection between a tracking ground station antenna and a LEO spacecraft.

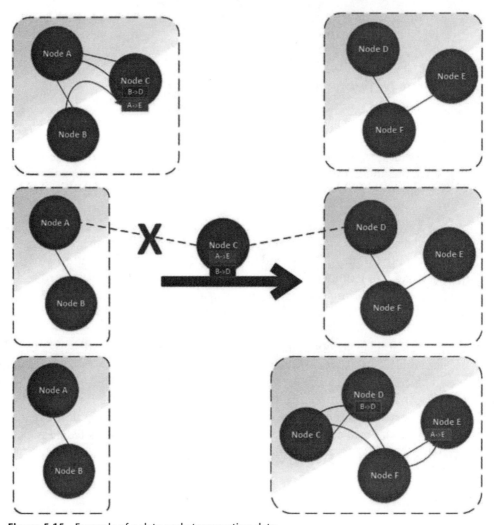

Figure 5.15 Example of a data mule transporting data.

5.2 Will a network of geosynchronous spacecraft provide a terrestrial internet-like experience for users on Earth? Explain how such a service would provide similar capabilities to those expected by a terrestrial internet connection. Explain how they would be different.

5.3 If you were developing a new computer network at Mars would you use the same infrastructure deployed on and around Earth? Why (or why not)?

5.4 Explain why a standard TCP/IP connection might not work if the latency between TCP session endpoints is too large. What modifications to TCP would need to be made to force it to work in this situation? How would those modifications affect the performance of TCP is more traditional operating cases?

5.5 Provide an example of a terrestrial internet connection that does not support bidirectionality. How is this example the same as the bidirectionality encountered in the SSI? How is it different?

5.6 Identify two ways in which coordinating deep space instrument observations may be different than coordinating near-Earth instrument observations. How might different networking technologies help provide a common way to operate both the deep space and near-Earth scenarios?

5.7 Identify and explain at least two differences between a terrestrial distributed sensor web and a near-Earth constellation of spacecraft that require different networking features.

5.8 Does the use of optical communications change the features needed of a space network? Why or why not?

5.9 Explain three limitations of using TCP/IP in an ad-hoc network consisting of data mules. Are these limitations addressed if the network uses DTN instead of TCP?

References

[1] Mai, T., "Tracking and Data Relay Satellite (TDRS) Third Generation Capabilities," *NASA Third-Generation TDRS*, NASA, May 1, 2015, www.nasa.gov/directorates/heo/scan/services/networks/tdrs_third_gen.

[2] Dunbar, B., "Tracking and Data Relay Satellite (TDRS) Fleet," NASA, May 1, 2015, www.nasa.gov/directorates/heo/scan/services/networks/tdrs_fleet.

[3] "Space Network Users Guide (SNUG), Revision 10," *Goddard Space Flight Center Document Number 450-SNUG*, NASA Goddard Space Flight Center, Greenbelt, MD, August 3, 2012.

[4] Cerf, V., et. al. "Delay-Tolerant Network Architecture: The Evolving Interplanetary Internet," draft-irtf-ipnrg-arch-01. txt, work in progress, 2002.

[5] Burleigh, S., et. al. "Delay-Tolerant Networking: An Approach to Interplanetary Internet," *IEEE Communications Magazine*, Vol. 41, No. 6, 2003, pp. 128–136.

[6] Delin, K., et. al., "Environmental Studies with the Sensor Web: Principles and Practice," *Sensors*, Vol. 5, Nos. 1–2, pp. 103–117.

[7] Wang, B., "SpaceX Low Latency Starlink Satellite Network Will Be Massively Profitable," NextBigFuture.com, December 13, 2018, www.nextbigfuture.com/2018/11/spacex-low-latency-starlink-satellite-network-will-be-massively-profitable.html.

[8] "Laser Communications Offer Promise," *SIGNAL Magazine*, Armed Forces Communications and Electronics Association International (AFCEA), November 9, 2018, www.afcea.org/content/laser-communications-offer-promise.

[9] "Start of Service for Europe's Space Data Highway," European Data Relay System, European Space Agency, November 23, 2016, www.esa.int/Our_Activities/Telecommunications_Integrated_Applications/EDRS/Start_of_service_for_Europe_s_SpaceDataHighway.

# CHAPTER 6

The Delay-/Disruption-Tolerant Networking Architecture

The obstacles to data delivery inherent in challenged networking environments are so complex that no single, magical protocol or application always guarantees message delivery. Constructing mechanisms for end-to-end delivery in challenged environments requires a balance of node placement, node processing capabilities, protocol engineering, and operational concepts for networking applications. DTN has emerged as one way to implement services in challenged networking environments. As an architecture, DTN node behaviors, new protocols, and has implications on the proper way to build networking applications. The remainder of this chapter discusses the ways in which the DTN architecture helps to overcome networking challenges.

6.1 Motivations for a Tolerant Network

The concept of a DTN was formalized by the IRTF in 2007 as RFC4838, titled "Delay-Tolerant Networking Architecture" [1]. This work was motivated by the observation that successful IP networks all make assumptions about their operating environment and that these assumptions may not be valid in many emerging challenged networking environments. If that is the case, then existing networking architectures may need to be updated so as to make fewer assumptions about their environments.

6.2 Assumptions Made by the Terrestrial Internet

RFC4838 lists several assumptions made by terrestrial networking. Among these assumptions are the need for an end-to-end path, timely feedback mechanisms, small, infrequent data losses, homogenous protocol support, and in-band performance tuning. Each of these assumptions is explained in this section, along with a description of why challenged networks violate these assumptions.

RFC4838 lists other motivations that are less impactful to the application layer, such as the assumption that endpoint-based security mechanisms are enough, that packet-switching is sensical, and that single route selection (or federating multiple discrete routes) will achieve desired performance. While these assumptions are also tenuous in challenged networking scenarios, they have less immediate impact on the design and implementation of delay-tolerant architectures and applications.

6.2.1 Path Existence

IP networks assume that an end-to-end path exists between a message source and destination. If the end-to-end path is disrupted at any time, the communication session may be lost in an IP network. This is certainly true in session-oriented protocols, such as TCP, but is also a consideration in session-less protocols, such as UDP when networking applications use acknowledgment and timeout schemes on top of UDP. This assumption has become synonymous with the concept of networking. For example, attempting to send an e-mail from a cell phone while in a place with no internet connection (such as a tunnel) is seen as nonsensical; since there is no end-to-end path there is no concept that a network exists.

However, networks can, and do, exist independently of end-to-end paths through the network; a node can claim membership of a network even at times when it is disconnected from other parts of the network. In challenged networks, links between nodes come and go over time as part of the nominal functioning of the network. There may never be a time when links between all the nodes comprising a path are available in the network. Further, there may be times when a node receives a message and its next link is unavailable for some period. This circumstance is treated as a temporary error state by IP and, thus, IP-based networks can handle this case using simplified mechanisms (such as dropping traffic).

6.2.2 Timely, Reliable, Actionable Feedback

IP networks provide reliability by retransmitting data based on the feedback (or lack thereof) received from downstream nodes. Retransmission typically corrects for an error encountered by the data in transit when that error could not be recovered by a lower layer in the protocol stack. If feedback from downstream nodes either arrives too late or arrives intermittently (or not at all), then transmitting nodes have no basis for tuning retransmission and fixed timeout schemes rarely produce efficient results [2].

In challenged networks, there is no guarantee that communications between nodes are bidirectional. When bidirectional connectivity does exist, it may simply be simulated as cooperating unidirectional links, and as such, the data rates may be significantly asymmetric. In these cases, feedback from the next hop node may have no direct way of getting back to the transmitting node (for unidirectional links) or may struggle to provide feedback in time (for heavily asymmetric links). In cases where there is no instantaneous end-to-end path in a network, downstream nodes may not be reachable and therefore cannot provide feedback to transmitting nodes. For example, if a downstream node loses the ability to store new information because its buffers are full, it cannot communicate that information to an upstream receiver if it is temporarily disconnected from the network. Therefore, challenged networks do not guarantee timely, reliable, actionable feedback.

6.2.3 Small End-to-End Data Loss

IP-based networks have two mechanisms for dealing with data loss: self-correction at the receiver and overall data retransmission. Self-correction involves using other layers in the protocol stack (both above and below IP) to correct minor corruption in data by adding annotations to the data (such as parity bits) or by sending the data multiple times (erasure/fountain coding). For small corruptions, this works well and is a requirement to correct for minor transmission errors experienced by all wireless transmission media. Where there is significant data loss—such as when a node in a path leaves the network—IP has no self-correcting mechanism other than to resend the data. This can be done passively where a node somehow infers that its data was not received and then sends the data again, or actively where a node assumes that the data is likely to be lost and sends it multiple times (or to multiple destinations). Both mechanisms make the assumption the data loss in the network is small and/or infrequent.

In challenged networking environments, data loss may be a normal occurrence, and in severe cases, successful data transmission can be the exception, not the rule. Parity bits and oversampling cannot correct severe data corruption or lack of reception. In cases where data is delivered but the acknowledgment of that delivery is lost, retransmissions are wasteful. In cases where end-to-end path loss is probable, there is no reason to believe that an end-to-end retransmission has any greater chance of succeeding than the original transmission. Because challenged environments cannot guarantee an end-to-end path and may not provide timely feedback, it is likely that there may be significant end-to-end data loss even when the network is functioning as intended. A new approach to handling data loss is required beyond those mechanisms deployed in IP-based networks.

6.2.4 TCP/IP Ubiquity

IP-based networks are the de facto standard for terrestrial networking. Networking architects, commercial appliance vendors, and application developers all assume that routers and endpoints in the network support the TCP and IP protocols and their related functions. This may not be the case in challenged environments where performance or resource constraints prevent the use of abstraction layers such as TCP and IP.

6.2.5 Performance Abstraction

IP-based applications typically do not worry about performance requirements in the network layer. While there are multiple protocols for tuning networking performance for special applications—such as video streaming—these settings are often hidden from the application layer. There is an assumption that applications may produce information at the desired rate and lower layers can prioritize and buffer that information and provide back-pressure and blocking where necessary to manage end-to-end performance. This assumption extends to any type of closed-loop control exercised in an IP-based network where in-order streaming data delivery, security monitoring, and actuator control represent just a small sample of performance-based functions coordinated across multiple network nodes.

Since challenged networks cannot guarantee feedback, it becomes difficult or impossible for real-time performance tuning to occur in the network layer. In these environments, applications must either abandon closed-loop control mechanisms or find ways to exploit knowledge about the environment or the type of data communications at the application layer. This necessitates moving the burden of performance assessment to the application.

6.3 Architectures for DTNs

A delay-tolerant architecture (DTA) is often conceptualized as an overlay network federating multiple individual nodes and potentially heterogeneous networks into a single system for passing messages. This formulation is a pragmatic consequence of the current state of message passing in challenged environments where there are no long-standing examples of networking infrastructure: networks are typically incremental accumulations of nodes enabled more by postdeployment reconfiguration of in-situ devices that may never have been designed to function as a network node. Because the expense of deployment prevents densely populating challenged areas with nodes, and because standards for delay tolerance are only now emerging in network engineering communities, DTAs will remain overlay networks for the foreseeable future.

To be effective, a delay-tolerant overlay requires several unique features that are separate from the construction and maintenance of any of its constituent nodes or networks. A conceptual DTA overlay architecture is illustrated in Figure 6.1.

In this figure, a variety of individual nodes and individual networks are federated into a single delay-tolerant overlay network[1]. Because an overlay network in a superset of its constituent networks, a network architect must consider some additional concerns relating to network features, protocols, and identification in the overlay.

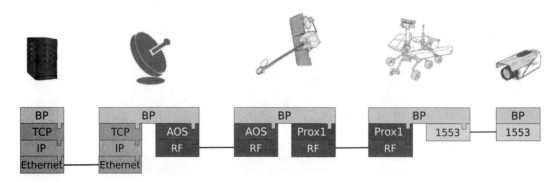

Figure 6.1 DTAs are often overlay architectures.

---

1. When a DTN is implemented directly as a single network, it can be viewed as a transparent overlay over that single network.

6.4 Delay-/Disruption-Tolerant Desirable Properties

The desirable properties of a DTA are listed in Table 6.1. Some of these properties may be satisfied by all, some, or none of the underlying networks participating in the overlay. Ensuring that these properties are consistently present in the overlay is the main goal of the delay-/disruption-tolerant network architect.

Most of these properties stem from the fact that, while IP-based networking is naturally ubiquitous in the terrestrial internet, IP does not gracefully expand into challenged environments. Forcing IP to exist in a DTA for compatibility with terrestrial networks may or may not be an efficient network design but keeping IP syntax does not mean that the services and algorithms colocated with IP services will work in a delayed or disrupted scenario. Fundamental network operations such as naming, routing, management, and security must be reimagined around the constraints of a challenged networking environment. Where possible, existing standards should be reconfigured and new standards should be proposed. Having individual nodes or individual networks provide one-off, incompatible solutions should be discouraged as it both prevents gradual accumulation of networking infrastructure in remote environments and places an unnecessary burden on network architects to reinvent solutions and rederive lessons learned available from other network deployments.

Table 6.1   Properties of a Delay-Tolerant Network

Property	Description
Overlay naming and addressing	The endpoints of a delay-tolerant network may exist in different address spaces. For example, when communicating data from a spacecraft to a mission operations center, computers in the mission operation center may use IPv4 addressing whereas there may be no IP stack at all on the spacecraft.
Late binding to physical resources	Just as overlay names may be different from the physical device names, the physical resources used for transmission might not be known in advance, especially in networks with mobile nodes.
Intelligent retransmission	Mechanisms for reducing source-to-destination retransmission must be employed to prevent network resources from being overwhelmed in cases where reliable transmission is often likely to fail.
End-to-end security	Security practices must span multiple, heterogeneous networks, to include networks that may not deploy IP and IP-based security mechanisms such as IPSec.
In-network storage	Store-and-forward operations require storage. While not every node in a network needs to be capable of persisting messages, there must be enough sufficiently resourced nodes to providing meaningful caching.
Routing	Overlay networks require routing solutions that span the individual networks they federate. Further, in cases where DTNs incorporate mobile nodes, routing algorithms must consider time-variant networking topologies that may not be present in the underlying routing algorithms of constituent networks.
Network management	Unlike the vast majority of terrestrial internet management, network management in a DTN must also be delay-tolerant. This requires the adoption of autonomy models and open-loop control algorithms.

6.5 DTN Protocols

A DTA is built from a variety of protocols working in coordination to provide necessary networking services. The set of DTN-related protocols mentioned in this section represent finalized standards, standards that are pending release from standards organizations, and proposed standards that are in draft form and being worked on by working groups in the context of a standards organization. Much as terrestrial IPs took decades to evolve, DTN protocols will continue to evolve. The two standards organizations representing the bulk of work in defining DTN protocols are IETF and CCSDS. The IETF standardizes protocols deployed across the Terrestrial Internet whereas the CCSDS standardizes several of the protocols deployed by national space agencies in their ground networks and flying on their spacecraft.

Table 6.2 lists a sampling of protocols in various stages of development, noting where they are in their maturity and how they provide services requires for a DTN to function.

Not all of these protocols are used equally in forming a DTN; some are needed all of the time whereas others may only be needed in certain types of networks, or portions of networks, or when attempting to use certain types of services. This is like the situation as it exists in the terrestrial internet where a few protocols are

Table 6.2 Various Protocols Related to DTN

Protocol	*State of Standard*	*Services Supported*	*Summary*
Bundle Protocol (BP)	BPv6 Finalized CCSDS Standard {bliste} BPv7 Pending IETF Standard.	Naming and addressing In-network storage Late binding	A transport protocol for messages across a DTN.
Licklider Transmission Protocol	Finalized CCSDS Standard	Intelligent retransmission	Reliable, single-hop protocol supporting intelligent retransmission timers for high signal propagation delay links.
Bundle Protocol Security	Pending IETF Standard	End-to-end security	Protocol for using secondary headers to sign and encrypt portions of a BP PDU.
Schedule-Aware Bundle Routing	Pending CCSDS Standard	Routing	Routing protocol for path planning in networks with scheduled node mobility.
Bundle-in-Bundle Encapsulation	Proposed IETF Standard	Security Routing	Protocol for tunneling BP PDUs.
Asynchronous Management Protocol	Proposed CCSDS Standard	Network management	Autonomy model associated with open-loop network management of nodes in a DTN.
Custody Transfer	Proposed IETF Standard	Intelligent retransmission In-network storage	Protocol for requesting, and implementing, storage of BP PDUs at waypoint nodes and making those waypoints responsible for future retransmissions of the PDU.

used all of the time and other protocols are only used where necessary. A graph of these internet dependencies typically has an hourglass shape showing convergence around IP, which is termed the thin waist of the internet. A similar graph can be made for a DTN and is illustrated (along with a traditional IP graph) in Figure 6.2.

As illustrated in Figure 6.2, the BP represents the thin waist of a DTN. It is the protocol that application layers can rely on to provide DTN services across a variety of lower layers.

6.6 Naming and Addressing

The term endpoint identifier (EID) is used generally in networks to identify data destinations (communications endpoints). In the context of DTNs, EIDs are represented as a Universal Resource Identifier (URI) [3]. The URI syntax represents identifiers as a tuple with the first element being a scheme name and the second element being a scheme-specific part. Scheme names associated with DTN networks include dtn and ipn [4], however, any valid URI can be used in this context. EIDs may resolve to URLs, IP addresses, spacecraft numbering schemes, or any other appropriate mechanism.

A single EID scheme should be used across the overlay to ensure that overlay protocols associated with security, routing, and management can function. However, this means that there must be a binding from overlay EIDs to the naming and addressing functions for individual physical nodes comprising the overlay. In certain cases, this may be straightforward, such as a one-to-one mapping of an IPv4 or IPv6 address to an EID. In other cases, the mapping may be more dynamic.

6.7 The BP Ecosystem

To fully understand the decomposition of roles and responsibilities necessary to create a DTA, it is important to take a more detailed view of the components forming the DTN. To the extent that a DTN is uniquely enabled by the BP, this means examining the components that allow the BP to work. These components are illustrated in Figure 6.3.

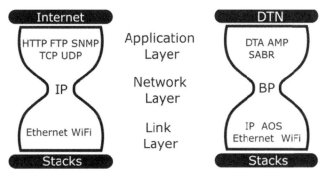

Figure 6.2 BP standardizes DTNs the way IP standardizes the terrestrial internet.

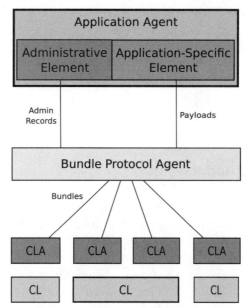

Figure 6.3 A BP agent exists as one component in a DTN ecosystem.

As shown in Figure 6.3, there are four major components to the BP stack: the applications using the BP, the BP protocol agent, convergence layer adapters, and the convergence layers they adapt. This section reviews each of these components as they enable (or are enabled) by the BP, starting with those components closest to the wire.

6.7.1 Convergence Layers

Convergence layers are those that deal with the final packetization, framing, and transmission of information through a medium. These convergence layers ultimately provide access to the underlying wired or wireless networks. The use of the word layer here is not meant to imply that there is a single protocol between a convergence layer adapter (CLA) and the network itself. Often, a single convergence layer will be represented by a stack of protocols whose top-most protocol interacts with the CLA and whose bottom-most layer interacts with the network. A common, terrestrial example of such a CL is the TCP/IP/Ethernet stack.

The capabilities of a given convergence layer may influence the types of BP services that are requested by the BP Agent. In cases where these services are known. For example, when a CL provides reliable fragmentation and re-assembly services, those services may not need to be invoked by the BP Agent.

6.7.2 Convergence Layer Adapters

Convergence layer adapters format information coming up from convergence layers into information that can be used by the BP Agent. This includes information about the received bundle, but also annotative data associated with the transmission and reception of that bundle. A CLA has three responsibilities, as follows:

1. Encapsulate a bundle from the BP Agent into one or more lower-level representations that can be understood by the convergence layer being adapted.
2. Decapsulate a bundle from one or more packets received from a convergence layer and pass it on to the BP Agent for processing.
3. Provide the BP Agent with any additional information required for the creation or processing of the bundle. This may include information relating to the capabilities of the adapted convergence layer, or annotative information received with a bundle. While the BP specification does not levy requirements on this type of information, implementations of CLAs do carry such annotative information as a necessity for optimizing performance.

6.7.3 The BPA

The BPA is software that created, processes, forwards, and delivers bundles to and from application and other BPAs. The behavior of the BPA is specified with the format of the BP in both the CCSDS [5] and the IETF [6, 7].

Despite being central to the concept of DTN, the BPA performs relatively few operations on its own. The BPA has three primary areas of responsibility: node registration, bundle processing, and bundle transmission/receipt.

A BPA is responsible for managing the mapping of EIDs to the node on which the BPA resides. Much like the mapping of IP addresses to MAC addresses, EIDs represent a logical mechanism to identify physical components. Multiple EIDs can be registered on a single node and the BPA must determine when an EID should be registered on the node, when an EID should be unregistered from the node, and whether these registrations should be suspended or not based on operation assessments.

Bundle processing by the BPA involves the inspection of the primary block and any extension blocks located with the bundle and acting on these blocks in accordance with their specification. For example, policy encoded in the bundle may stipulate that if the block cannot be processed by the BPA, that the BPA should remove the block from the bundle (or in some cases discard the entire bundle). Some extension blocks define networking information, such as the prior bundle node that transmitted the bundle. Others may include flow labels, security information, extended quality of service, or any of a myriad of potential protocol extensions. Just as this self-extension mechanism is an enabling feature of the BP, the processing of these extensions is an enabling feature of a BPA.

Transmission and receipt are fundamental concepts for any protocol agent, and the BPA is no different in this regard. Upon receiving a bundle from a CLA, the BPA must determine whether the bundle is destined for the current node or some other node and, if the current node represents the destination of the bundle, the BPA must deliver the node to (or through) the node's application agent (AA). Otherwise, the bundle must be queued for later transmission. Similarly, the BPA transmits bundles when informed to do so as a function of perceived available contact opportunities (either as given by a schedule or as determined by some opportunistic mechanism). BPAs transmit (a) previously received bundles for which this node is not the final destination, (b) new bundles sourced at this node, and (c) any previously transmitted bundle that it deems must be retransmitted.

6.7.4 Application Agent

The AA represents the user application(s) that use the BP Agent to communicate information to and from other AAs at other nodes in the network. The roles and responsibilities of such agents are, according to the BP specification, logically decomposed into two separate responsibilities: administration and application-specific processing. Depending on how a BP node is constructed, the AA may either be a single entity that provides a common interface to a multitude of user processes or may be combined with user processes. Either way, the AA represents the BPA entry point into the user space much as the CLA represents the BPA entry point into the network. For this section, we will not distinguish between the AA and the user processes that utilize (or are merged with) the AA.

The administrative role for an AA uses the concept of administrative records associated with bundle transmission. As covered later in this chapter, there are several instances in which a bundle may cause an administrative record to be generated and sent to an administrative node (which may be different from the source node for the bundle or the destination node for the bundle). Processing these administrative records forms a type of open-loop control for the AA, and it is the first clear signal that the BPA itself is a necessary but not sufficient mechanism to handle the end-to-end exchange of data in a DTN. However, administrative records alone cannot provide all of the information necessary for open-loop control. Besides the reality that administrative records might not be received in a reliable and timely way, attempting to build and maintain situational awareness for all administrative records for all bundles in the network quickly becomes an unmanageable task.

The application-specific processing role for the AA consists of the usual functions associated with user applications, such as data generation and consumption.

The role of decomposition into administrative processing and application-specific processing is not a clean boundary. Just as administrative review of the underlying DTN cannot be wholly accomplished through administrative records, the decision of what information to generate and to whom it should be sent, cannot be effectively accomplished in a DTN without some understanding of the state of the network.

6.8 Special Node Characteristics

The aforementioned set of AAs, BPAs, CLA, and CLs operate on a single node (and perhaps every node) in a network. To fully implement all of these features, there are several unique characteristics that these nodes must possess in order to completely perform their duties in a DTN. Several of these features may also be present in nodes that are not part of a DTN, but when operating within a DTN, these features may be enabling for store-and-forward message exchange.

These special node characteristics almost all involve some expenditure of resources in the design of the node or for its operation. As the links in a network become less reliable, the resources needed by the nodes on either end of those links must become more resourceful. That correlation should not be a surprise to the network architects and operators looking to build out these networks. However, the rationale for resourced nodes is more nuanced than that. Supporting a DTN architecture is not simply additive. If supporting a DTN simply meant that nodes must keep adding resources and processes on top of existing and unmodified resources and processes, then the cost of supporting such a network could start to outweigh the benefits of such a network. Instead, a DTN architecture must be implemented as a transformative process. Resources are added to a node while, simultaneously, removing other processes or resource expenditures from the node. In this way, the DTN architecture can start to pay for itself in that node resources may be applied to more efficient mechanisms in challenged communications conditions rather than used to support inefficient strategies for transmission/retransmission. The general discussion of the impact of delay-tolerance (in both networks and applications) as a driver for network architecture is addressed in much more detail later in this book.

6.8.1 Persistent Storage

No store-and-forward mechanism can exist without some mechanism for storage. Nodes within a network must provide some method for storing information beyond that which exists within transient buffers used to maintain a specific performance metric. Persistent storage, in this case, means any storage mechanism (or mechanisms) which can hold on to a bundle outside of the resources already allocated for bundle transmission and reception. Having resources that exist out-of-band from bundle transmission ensures that bundle processing and storage do not otherwise burden the reception and transmission of bundles at high rates. For example, the number of stored bundles on a node should not significantly alter the rate at which a node can receive bundles or transmit bundles.

Persistent storage does not imply any particular storage mechanism; in fact, it is likely that nodes in a DTN will use different storage media based on their supported data rates and resources. For example, high-powered servers representing a boundary between a highly available network and a DTN may use hard drives or commercial solid-state drives for persistent storage, whereas resource-constrained, embedded devices might directly write to a NAND flash with or without a file system. Other types of devices may implement persistent storage as a special allocation of ring buffers in memory or on custom-build (or soft-core) processors. Nodes may have a single layer of storage or multiple layers of storage with a small amount of fast cache access and a slower, larger backing store. Nodes may store bundles in contiguous sets or may parse them in ways that make their lookup and reassembly more resource efficient.

It is important to note the myriad of ways in which nodes can provide persistent storage solutions within a DTN, because the requirement for storage in these networks is not itself an impossible task in the context of a resource-constrained system. Very small, embedded systems with strict power constraints may actually save power by reducing retransmissions in the network for the price of the power needed to support an additional (or larger) storage mechanism.

6.8.2 Late Binding

Late binding, in this context, refers to the ability of one node to stand in for another node as a function of its physical link abilities as well as its processing and network capabilities. In several next-generation networking architectures, the concept of targeting data to a single node has become obsolete, as it assumes that there remains a one-to-one relationship between a node identifier (typically an IP address) and a user application relying on that data. In cases of terrestrial networking where nodes are not typically mobile, that mapping may be valid. However, in a challenged network where nodes may have constant mobility, it is possible that both waypoint nodes and sometimes destination nodes may be interchangeable.

A common example of that interchangeability can be found in data mules that provide regular data exfiltration opportunities for sensor nodes. In this case, a particular sensor node does not particularly care which unmanned aerial vehicle or LEO satellite takes its data, as long as the data can get off of the sensing platform. In this case, the sensing node would not necessarily need (or want) to broadcast data addressed to multiple nodes in the network and, depending on the situation, the use of multicast broadcasts may be discouraged.

In cases where EIDs are not treated as multicast addresses, the BPA remains able to register nodes with an EID and to remove or suspend that registration. In these cases, nodes may be late-bound to EIDs as a matter of policy or in reaction to events.

6.8.3 Multiple Convergence Layers

Multiple convergence layers can refer to different protocol stacks running over the same physical link, or difference protocol stacks running over different physical links. There are multiple ways in which a node could represent multiple convergence layers. We represent these options in terms of a mapping from the highest level of a supported protocol stack to the physical link carrying information from that stack. Nodes can support a one-to-many, many-to-many, and many-to-one relationship, as illustrated in Figure 6.4.

6.8.3.1 Many-to-One

Multiple convergence layers using a single physical link provide different services over that link. For example, a node supporting a Wi-Fi or Ethernet connection

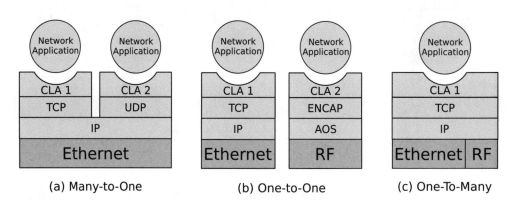

Figure 6.4 There is a complex mapping of convergence layers to physical links.

could use either a TCP/IP or a UDP/IP stack and since the top-most layer of these stacks are different, they would be seen as different convergence layers requiring their own convergence layer adapters: a TCP adapter and a UDP adapter. Much like any other network application, the decision to use TCP or UDP is based on the types of services require for the data being transferred.

6.8.3.2 Many-to-Many

Multiple convergence layers for multiple physical links reflects the case where nodes participate in a heterogeneous network and support multiple networking connections (typically via a software defined radio or multiple radios on a device). In these instances, it can be the case that the node is in multiple contact with multiple radios at the same time, but more often it is likely that the node uses networking diversity to join different parts of a network (or different networks) at different times.

6.8.3.3 One-to-Many

A single convergence layer supporting multiple physical links reflects the case where the node itself wishes to manage network access as a function of technical or administrative policy. A primary example of such an arrangement being a node which would provide a TCP stack that sits over Bluetooth, Wi-Fi, Wi-Max, or other mechanisms. The selection of which physical link carries the TCP traffic would be based on user policy, availability, data rates, and other methods that would otherwise be considered transparent to the user.

The fact that nodes support multiple convergence layers and, at times, multiple physical links, indicates that neither the AA, BPA, nor CLA have complete control over how data eventually gets to the network. In many cases there are multiple control loops associated with retransmission, reassembly, and health and status data being communicated. The abilities of these multiple convergence layers are important to exploit known architectural details in portions of challenged networks that may not exist in other portions of the network (or on other nodes in the network). This places a large and often impossible burden on the CLA and BPA to determine how to best manage this complexity.

6.9 Summary

There is no single solution to the general problem of exchange data in challenging networking environments. Because these challenges are distributed across application, network, and link layers, solutions to these problems must similarly be distributed across multiple architectural layers. One proposed architecture for these environments is the delay-/disruption-tolerant architecture. This architecture was a natural convergence of the output of multiple research and engineering efforts by industry, government, and academia.

The IP is the converging protocol for several terrestrial internet applications and operational concepts, and BP serves a similar role for DTAs. Because networks in challenged environments are often built by federating multiple discrete networkable assets, BP is often implemented as an overlay protocol relying on convergence layers and convergence layer adapters. Further, to fully implement a DTA, nodes in the network must support additional features that are not always needed in IP networks.

6.10 Problems

6.1 Do you agree with the statement that the functioning of IP-based networks relies on critical assumptions? What assumptions might be missing from this list?

6.2 Provide an example of an IP-based networking service that functions without needing an end-to-end path. How is data exchanged in this circumstance, and does this function perform activities similar to BP?

6.3 Describe a scenario where feedback could be sent back to a sender transmitting over a unidirectional link in a network. Would this feedback mechanism provide a reliable indication of message delivery? Why or why not?

6.4 Given a networking path of five hops starting from Node N0 through nodes N1, N2, N3, and N4 with a final destination of N5, with each link having a 20% chance of dropping a message, calculate the number of times a message would need to be transmitted to achieve at least a 90% delivery probability in the following three scenarios: (a) that retransmissions always occur at the source node, (b) that custody is accepted by even nodes in the network, and (c) that custody can be accepted by all nodes in the network.

6.5 Give an example of late-binding to physical resources in a delay-tolerant architecture. Provide a mechanism for mapping a logical endpoint to a physical node identifier. Use either DTN or IPN schemes and provide syntactically correct EIDs.

6.6 Provide three reasons why a session-based, query-response based network management protocol, such as the SNMPv3, would not operate efficiently in a delay-tolerant network.

6.7 Calculate the protocol overhead associated with the following three different protocol stacks, assuming that each protocol layer includes all optional fields with a payload size of 1,024 bytes: (a) TCP/IP/Ethernet, (b) BP/TCP/IP/Ethernet, and (c) BP/Ethernet.

6.8 Given a BPA connected to 1 Gbps link that received 1,000 PDUs per second resulting and a consistent 30% link utilization, how much storage would be needed by the BPA to store these PDUs for two seconds? Assuming no additional overhead associated with PDU storage, what would the write speed of the storage need to be in order to not fall behind?

6.9 Assume that a BPA is running on an iPhone X. How many convergence layer adapters would be necessary to provide access to all of the radios on the device? Diagram the mapping over CLAs to radios and the CL stacks between them.

References

[1] Cerf, V., et. al. "RFC 4838," *Delay-Tolerant Networking Architecture*, IRTF DTN Research Group, April 2007.

[2] Jain, R., "Divergence of Timeout Algorithms for Packet Retransmissions," arXiv preprint cs/9809097, 1998.

[3] Berners-Lee, T., R. Fielding, and L. Masinter, "RFC 3986, Uniform Resource Identifier (URI): Generic Syntax," URL: http://www. faqs. org/rfcs/rfc3986.html, 2005.

[4] Clare, L., S. Burleigh, and K. Scott, "Endpoint Naming for Space Delay/Disruption Tolerant Networking," *2010 IEEE Aerospace Conference*, IEEE, 2010.

[5] *CCSDS Bundle Protocol Specification, Recommendation for Space Data System Standards (Blue Book)*, CCSDS 734.2-B-1, Washington, DC: CCSDS, September 2015.

[6] Scott, K., and S. Burleigh, B*undle Protocol Specification RFC 5050*, Reston, VA: ISOC, November 2007.

[7] Burleigh S., K. Fall, and E. Birrane, "Bundle Protocol Version 7," Working Draft, August 2019, IETF Secretariat, http://www.ietf.org/internet-drafts/draft-ietf-dtn-bpbis-14.txt.

CHAPTER 7
Patience on the Wire: The DTN BP

The BP seeks to be to DTN what IP is to the terrestrial internet: a standardization of best practices and a consolidation of APIs that allows for interoperability. Because the BP will likely be the protocol used to communicate delay-/disruption-tolerant application data, it is important to take a closer look at the specifics of the protocol beyond the roles and responsibilities of a BPA. This chapter discusses the overall goals of the protocol, the structure of the bundle as the BP protocol data unit (PDU), and those elements of the protocol that provide beneficial features when operating within a challenged networking environment. Finally, we compare BP with the features provided by IP and address the fundamental question: Is BP enough to claim delay/disruption tolerance?

7.1 Protocol Goals

The BP provides a standardized mechanism for store-and-forward semantics and behaviors in a network. In a store-and-forward model, messages received by a node in the network are expected to store a message until an appropriate future transmission time or some other prearranged condition is met. In the most literal sense, all networks are store-and-forward because messages exist in transmission buffers and memory while pending routing and forwarding decisions or simply while awaiting their turn on the wire. In fact, properly tuned caches can decrease latencies in a network [1]. However, in a well-tuned resourced networking environment, this type of storage is at best a necessary obstacle to data communications. Network appliances in resourced environments seek to store this data for as short a time as possible, with this time (delay) often measured in milliseconds. Short storage times are desired by network designers and hardware manufacturers as they increase data throughput and storage is often upper-bounded by the practical concern as to how much space is needed and available to receive high-rate data.

For example, a saturated 1-Gbps link would need almost 140 MB of buffer space to store incoming messages (without additional annotative and/or management data) for one second—a relative eternity in the terrestrial internet. BP traffic, however, may request storage for multiple seconds, minutes, hours, or days. Handling this storage requirement requires additional protocol overhead, bundle processing, and storage management. The features of BP must justify the additional expense and complexity.

7.2 The Case for BP Store and Forward

Data flow in a terrestrial network is both unforgiving and chaotic. Network collisions happen dozens of times per second even over wired links, and messages are only allowed to rest in buffers for milliseconds before they are transmitted off node or summarily deleted. This occurs against a backdrop of quality of service and service-level agreements enforced through policies of message deletion and retransmission. Each of these events occur on a link-by-link basis, multiplied many times over by the number of simultaneously operating independent links throughout the network. The high-rate data exchange so often taken for granted in the terrestrial internet is the result of a massive and, at times, manually tuned infrastructure. As a consequence, this infrastructure is necessarily fragile when assumptions about connectedness and availability fail.

The strategy of keeping things moving by deleting messages that get in the way (such as those messages of lower priority or whose next hop is unreachable or unresolvable) is pragmatic from the point of view of a forwarding node. However, each deleted message represents a potential timeout-and-retransmission event from the perspective of the message's sender. Retransmitting a low-priority message results in that message consuming multiple networking resources multiple times; a counterintuitive mapping of resources to low-priority data, especially along a multihop path. Managing this situation requires a balance between deleting messages and suppressing wasteful retransmissions. This is a difficult balance to achieve, much less optimize—particularly along a complete end-to-end path—and most networks avoid the problem by keeping overall network utilization low enough that retransmissions do not impact performance. For example, it is recommended that an Ethernet network operate at no more than 70% networking utilization because at least 30% of available bandwidth may be needed to handle coordination and retransmission information [2, 3]. This approach is pragmatic, but wasteful in terms of resources throughout the network, and becomes increasingly burdensome as the end-to-end number of hops in a path increases from source to destination. Adopting new techniques such as flow utilization can increase individual burst times in the network, but most network utilization is kept well below 70% [4].

There is a fundamental scalability challenge with this overprovisioning approach: as networks grow in size and complexity, the amount of data present in the network is increasing faster than the ability to provision additional networking resources. This situation exists for two reasons. First, the number of data producers—and the amount of data they produce—is increasing. Second, we are instrumenting parts of our world and our solar system where there is little ability to establish and/or grow a networking infrastructure. In these contexts, dropping a message at a given node is discouraged because it may have consumed significant resources to get the message up to that node in the first place. Similarly, transmitting the message as soon as possible may not be practical because there may be no resources at the node for transmission (as a function of power, link bandwidth, antenna pointing, etc.) or a link to the next node may not be immediately available either as a result of scheduling, failure, or visibility (and other physics challenges resulting from geometry, topology, or as is the case in space communication, the motion of spacecraft and planets, etc.). The only remaining option is for a node to hold on to the message hoping for more information later. The BP provides a standardized way of doing exactly that.

7.3 Services Unique to BP

The BP provides three major services: store-and-forward operation, self-annotation, and layer-agnostic operation. There are other syntactic differences between BP and other protocols such as IP, and intriguing networking application possibilities based on BP that are explored later in this chapter and in this book. However, all of these possibilities begin with these three motivating features of the BP:

1. Store-and-forward operation. The fundamental feature of the BP is its expectation that bundles be persistently stored—when necessary and possible—by BPAs in the network. The behavior, signaling, and definitions of how this storage should be accomplished and communicated are a significant portion of the BP specification.
2. Self-annotation. The bundle provides a self-extension mechanism by which secondary headers may be added to supplement the payload. This mechanism allows bundles to be effectively communicated without the need for establishing (and maintaining) end-to-end sessions to maintain shared understanding between message sources and destinations.
3. Layer-agnostic operation. BP does not presume any particular underlying networking layer (and, therefore, does not assume that such layers provide specific abilities). BP may run over TCP, UDP, raw IP, or lower-layer frames. Some researchers have proposed running BP using USB drives as an underlying transport. While the concept of protocol layering is to be agnostic of other layers in a protocol stack, it is often the case that the practical implementation and optimization of protocols are based on assumed features occurring elsewhere in the stack.

7.4 Protocol Layering Considerations

The layer-agnostic functioning of the BP is important because the position of BP is not fixed within a standard protocol stack ordering. BP is designed to operate on a variety of platforms, including those with heavy resource constraints, and therefore has very few hard-coded assumptions relating to the capabilities of its underlying layers. Because BP makes fewer assumptions, it becomes possible to build convergence layer adapters for a variety of networking scenarios. Some options of BP in various layers of the protocol stack are illustrated in Figure 7.1.

An excellent example of the flexibility of BP regarding its position in a protocol layering is the ability to use a BP convergence layer adapter for the BPA itself. A conceptual specification for performing exactly this type of function is currently being drafted in the IETF DTNWG under the auspices of "Bundle in Bundle Encapsulation" [5]. In BIBE, a bundle may be fully encapsulated inside of another encapsulating bundle for the purposes of tunneling, security, or other mechanisms.

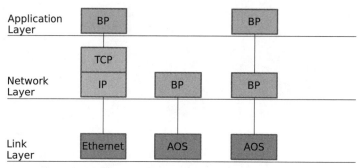

Figure 7.1 BP implemented at various layers in a networking stack.

There are, of course, circumstances where this type of encapsulation is beneficial and times when it is not and cases where tunneling or security is more effectively implemented as part of the network application design patterns.

There is likely no better example of BP's lack of assumptions on its lower layers than the example of BP having a CLA that itself sits over BP. For the remainder of this chapter, there is no assumption associated with which layer implements BP or with the functions provided to it by any underlying CL or CLA.

7.4.1 Versions of the BP

While this book focuses on the concepts of DTN architectures in general—no matter their specific implementation—it is worth noting that the fundamental BP is undergoing change and that the proposed changes will likely result in challenges that will need to be addressed by network designers and DTN application developers. The change at hand is the revision within the IETF of the BP specification (RFC 5050) from version 6 (BPv6) to version 7 (BPv7). In the process of revising the specification several decisions have been made as to how to implement existing and new functionality that are not backward compatible from BPv7 to BPv6. The implication of this incompatibility on the network and application developer is that in a mixed version environment care must be taken as to which version of BP a particular bundle and bundle agent is based on, and if interoperability is a goal, translation services must be provided within the network.

The significant changes from BPv6 to BPv7 identified in the draft DTNWG specification [6] are as follows:

1. A distinction is made between the concepts of transmission and forwarding of bundles;
2. The mechanics of custody transfer in BPv7 are addressed through the bundle-in-bundle encapsulation (BIBE) specification;
3. The addition of a concept of node ID that is distinct from the concept of EID;
4. The addition of a new method for encoding endpoint and node IDs in a bundle and removing the concept of the dictionary and use of CBHE for header compression;
5. Restructuring of the primary block and adding additional CRC and inventory fields;

6. Adding optional CRCs to nonprimary blocks;
7. Addition of block ID number to canonical block format supporting streamlined BP Security (BPSEC);
8. Addition of several new extension blocks including the bundle age extension block, previous node ID extension block, flow label block, and the hop count extension block;
9. Removal of quality of service markings;
10. Changing the custom encoding scheme to the more standard CBOR encoding scheme.

Regardless of the specific BP version implemented, the underlying principles and behaviors of BP and DTN architectures remain the same. This book is about designing and implementing applications and architectures that provide delay/disruption tolerance using the BP. While the details of exactly the format of the primary block and the behaviors of certain extension blocks will differ between BPv6 and BPv7 the underlying design and implementation of a DTN architecture and bundle aware applications will remain the same.

Therefore, when discussing specifics of the BP, the book assumes BPv7 as the latest incarnation of the standard, because the concepts, philosophies, and utilization of delay-tolerant concepts remains the same regardless of whether a BPv6 or BPv7 syntax is used to express them.

7.5 Bundle Structure

The structure of the bundle reflects the need for the BP to not rely on timely, stateful data exchange. Similar to other protocols, a bundle is comprised of a primary header and a payload. Additionally, bundles may include secondary headers and those secondary headers may be added at multiple points along a path and may hold a variety of information types. Each logical section of a bundle (primary header, secondary header, payload) is termed a block. In that sense, a bundle is a bundle of blocks (see Figure 7.2) that are treated as an atomic n-tuple of information.

The blocks within a bundle contain both optional and variable-length fields. Flags in blocks contain information that indicate the presence or absence of optional fields. Variable length fields can be self-delineating or a combination of length-value pairs. In all cases the design goal is the same: increased node processing to enable more compact transmissions.

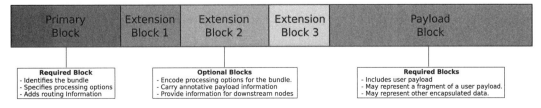

Figure 7.2 A BP PDU is a bundle of blocks.

Table 7.1 Common Block Fields for the Payload and Extension Block Types

Item	Description
Type	This variable-length, self-delineating value describes the type of block that follows. Every possible block in a bundle is assigned a block type and these type assignments are moderated through the IANA.
Number	This variable-length, self-delineating value serves as a differentiating mechanism for multiple blocks in a bundle of the same type. No two blocks in a bundle may have the same block type and block number. Other than as a differentiation mechanism, no other meaning should be inferred from a block number. Blocks of different types may have the same block number.
Control flags	This bit field denotes which policy decisions must (or must not) be applied to this block when processing the bundle. These decisions include options such as removing the block from the bundle (or deleting the bundle) if the receiving BPA does not understand the block type.
CRC type	This mandatory field describes the type of CRC that has been computed for the block. When set to 0, this indicates that there is no CRC on the block. The decision to use CRC values is based on assumptions relating to the underlying convergence layers and whether other integrity mechanisms are used in the bundle.
Type-specific data	This is a variable-length, self-delineated field that contains the serialization of block-specific information.
CRC	This optional field includes a CRC calculated over the entire block (to include the CRC field itself, which will have an initial value of all 0s).

With the exception of the primary block, all other blocks in a bundle share a common format, captured in Table 7.1.

7.6 The Primary Block

Similar to primary headers in other protocols, the bundle primary block contains information necessary to uniquely identify the bundle and provide information which may be helpful for routing/forwarding decisions. Unlike some other protocol primary headers, the information in the BP primary block may be superseded or augmented by information in extension blocks. For example, a primary block may specify a bundle destination, but an extension block may also specify a flow label for the bundle which could take priority based on the processing rules for that extension block. The fields of the primary block are defined in [6] and in the following sections.

7.6.1 Version

This variable-length, self-delineating value describes the version of BP used to create this bundle. At the time of this writing, v6 and v7 are valid BP versions with v6 standardized by the CCSDS and v7 being standardized by the IETF.

7.6.2 Processing Control Flags

This variable-length, self-delineating value contains a series of flags (and reserved for future use allocations) that describe how a BPA should process bundles in certain circumstances. This includes when to send status reports, what types of blocks should be expected in the bundle, whether the bundle must be acknowledged, whether the bundle is a fragment, and other housekeeping items.

7.6.3 Cyclic Redundancy Check Type

This mandatory field describes the type of cyclic redundancy check (CRC) [7, 8] that has been computed for the primary block. When set to 0, this indicates that there is no CRC on the block. The decision to use CRC values is based on assumptions relating to the underlying convergence layers and whether other integrity mechanisms are used in the bundle. Based on transmission characteristics and the environment, different CRC algorithms can be used in bundles.

7.6.4 Destination EID

This variable-length, self-delineating value captures the EID to which the bundle should be delivered. If the receiving node is registered with this EID, then the BPA must deliver the bundle to the AA at this node. Otherwise, the BPA must determine a next bundle hop and forward the bundle as appropriate.

7.6.5 Source Node ID

This variable-length, self-delineating value captures the node that produced the bundle. A node ID is simply an EID that uniquely describes a BPA node. The node itself is differentiated from services that run on the node. In certain cases, this identifier can be set to a NULL value, in which case the bundle is considered anonymous and any processing that requires knowledge of the bundle source must infer that information from other mechanisms such as extension blocks or out-of-band configuration.

7.6.6 Report-To EID

This variable-length, self-delineating value stores the EID to which reports generated by this bundle should be sent. When applications must rely on open-loop control mechanisms, one design pattern is to have status information routed to a central status repository rather than having individual statuses communicated back to a myriad of source nodes.

7.6.7 Creation Timestamp

This pair of variable-length, self-delineating unsigned integers represent the time that the bundle was created. The first integer represents the bundle's creation time while the second represents a sequence number used to disambiguate multiple bundles generated at the same time.

Simply increasing the resolution of the timestamp (for example from seconds to milliseconds to nanoseconds) both increases the size of the timestamp and may not add fidelity for constrained nodes that are unable to maintain clock accuracy and must approximate finer-grained time values.

7.6.8 Lifetime

This variable-length, self-delineating unsigned integer represents the number of microseconds past the bundle's creation time before which the bundle should be considered expired. Expiration is calculated by user applications and denotes the time at which either the bundle information is no longer considered useful or the time after which the bundle is consuming too many resources relative to the value of its information or some other metric.

7.6.9 Fragment Offset

If the bundle is marked as a fragment, then this variable-length, self-delineating field contains the offset of this fragment in the eventually assembled whole bundle. If the bundle is not marked as a fragment, then this field is omitted from the primary block.

7.6.10 CRC Field

This optional field includes a CRC calculated over the entire block (to include the CRC field itself, which will have an initial value of all 0s).

7.7 The Payload Block

The payload block is an extension block whose block type is set to the IANA-assigned value for the payload, and whose block-type-specific data is exactly the bundle payload. There can be only one payload block in a given bundle, and as such, the payload conceptually contains the primary information whose conveyance is the reason the bundle exists in the network. All other information in the bundle exists to support the delivery of the payload. We say conceptually in this instance because there are proposed designs that consider including payload summary data outside of the payload block to aid in the development of delay-tolerant applications.

7.8 Extension Blocks

The term extension block refers to any block within a bundle that is neither the primary block nor the payload block. Extension blocks provide an extensibility mechanism for BP and allow the BP to work in challenged environments that either cannot maintain or establish a communications session. Extension blocks and communications sessions both attempt to solve similar problems in dissimilar ways and therefore the mechanisms are conceptually related.

The purpose of a communications session is to hold, at endpoints along a path, synchronized state information that is helpful for end-to-end data exchange. This may include information such as session keys, timeout values, congestion measurements, back-pressure signals, and flow associations. When nodes are constantly joining and leaving a network, and when bundles are stored at nodes for long periods of time, these kinds of session data expire before they can be useful. The alternative to session-state synchronization at endpoints is to carry relevant information with the bundle itself. Extension blocks provide the means to accomplish this. Therefore, the original concept of extension blocks represents annotations associated with the bundle's payload or the bundle itself. By not combining annotation with the payload, nodes in a network may observe, process, and update annotative data without needing to understand, access, or modify the bundle payload.

Extension blocks also have utility beyond carrying session-like information. In some cases, extension blocks can be used to implement a protocol colocated with the bundle itself. BPAs may alter the bundle, or alter individual extension blocks, based on the contents of a particular extension block. For example, an extension block might carry link-state information with a bundle that can be used to configure local routing and forwarding decisions for each node visited along the bundle's journey [9]. In this way, such an extension block would act as a delay-tolerant link state dissemination protocol colocated with the bundles over that link. Various ways in which extension blocks can be used is illustrated in Figure 7.3.

The format of the block-type-specific data fields of an extension block, and the behavior of BPAs when encountering a specific type of extension block, are formally specified, either in their own documents (such as the case with bundle security [10]), or coupled with other protocol specifications. The BPv7 specification [6] lists three extension blocks that provide annotative information helpful for bundle processing: the previous hop, bundle age, and hop count blocks.

An example of the use of extension blocks to help with per-bundle processing is the previous node block (PNB), which can be added to a bundle by a node prior to transmission so that a future receiving node can identify the last BPA to have forwarded the bundle. The PNB is an excellent example of the kind of bundle annotations that are important for certain types of validation, particularly in the absence of a communication session. A BPA on a node has no other readily apparent way of identifying the previous BPA responsible for sending a locally received bundle. Not only might that information not be passed from a CL through a CLA, but the previous physical hop might not have been a BPA to begin with. Figure 7.4 illustrates the

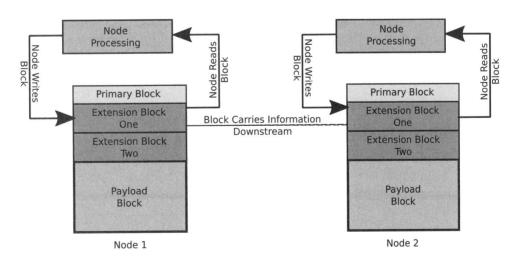

Figure 7.3 Extension blocks annotate payloads and configure BPAs.

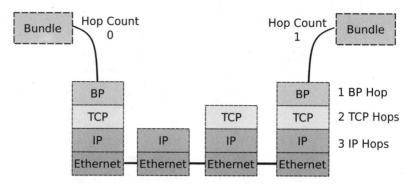

Figure 7.4 A previous hop in an overlay versus in a physical network.

situation where the previous physical hop is not the previous bundle hop. In this case there are multiple hops across independent network links that must be completed to result in a single hop between BPAs. This is an example of per-bundle-hop annotation. The PNB is expected to be removed by a receiving BPA and, perhaps, replaced with a new PNB by the BPA before further transmitting it again.

Similar to the PNB, the bundle age and hop count blocks provide annotative information for the bundle that otherwise is unavailable from CLAs, control-channel traffic, or session state. The bundle age block contains the number of microseconds that have elapsed between when a bundle was first created and when it was last forwarded by the PNB. This is an attempt to help applications account for long signal propagation delays and periods when a bundle has been stored at a node. The hop count block counts the number of hops taken by a bundle through BPAs along the path to inform BPA decisions on when a bundle has been forwarded too many times in a network.

7.9 BP-Enabled Concepts

As of this writing, several other extension block types are in active discussion in various DTN-related working groups. A compendium of the syntaxes and standards for these blocks is well beyond the scope of this chapter and this book. However, looking at the kinds of protocol behaviors enabled by these blocks is precisely the type of information that is useful for network architects as they plan for the deployments of DTNs, and network programmers writing the applications that run in those DTNs. Readers interested in learning more about additional extension blocks are encouraged to look to the IETF DTNWG.

7.9.1 Application Annotations

As previously mentioned, extension blocks are primarily intended to capture session and other state information associated with a bundle to help it on its journey through a challenged network. However, applications may also provide information to be included in a bundle, separate from the payload, to help with the disposition of the payload (rather than the bundle) through the network. These annotations can provide insight into the contents of the payload without providing access to the payload itself, or include secondary information that should be coupled with the payload but perhaps updated separately from or without modifying the payload.

For example, an application may wish to attach one or more names to the data represented in a payload. In doing so, the bundle could be routed, stored, and otherwise processed based on the name of the data without a need to parse the actual data stored in the bundle payload. In this way, behavior such as named-data networking [11] or information-centric networking [12] heuristics could be employed in a BP network.

Alternatively, applications at a source node (or at waypoint nodes) could provide information useful for the destination of a bundle, such as the node ID, time, or other identifying information regarding the nodes that have forwarded or otherwise handled the bundle over time. This would provide a functionality similar to a delay-tolerant traceroute function, where a destination node would receive a bundle with either multiple location extension blocks or a single route path extension block that has been annotated over time and through the network.

Finally, some annotations may be used to provide not just alternative identification for the payload, but a summary of the contents of the payload. For example, a sensor node may construct a payload with a series of high-fidelity sensor readings or data that requires significant post-processing for interpretation in such a way that the payload should be protected from inspection by anyone except the destination node. However, a summary, simple average, or quick look of the sensed data could also be included in an extension block with the bundle for waypoint nodes. By bundling user payload and summary information together there is no concern that the two pieces of data would be separated in the network, or alternatively, interim recipients in the network path can make use of the quick look information while the detailed data is in-transit to its eventual post-processing point. These concepts are illustrated in Figure 7.5

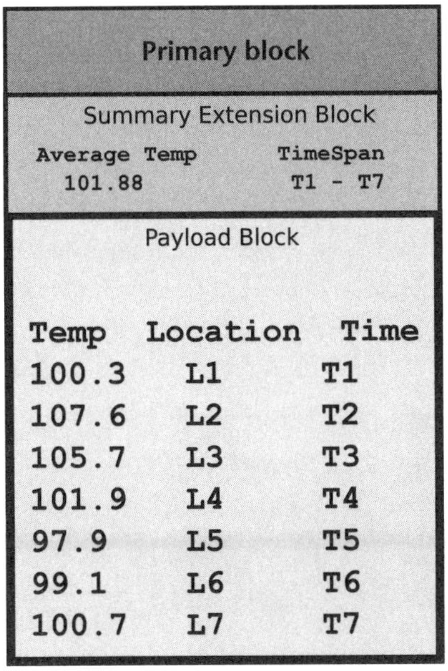

Primary block		
Summary Extension Block		
Average Temp 101.88		TimeSpan T1 - T7
Payload Block		
Temp	Location	Time
100.3	L1	T1
107.6	L2	T2
105.7	L3	T3
101.9	L4	T4
97.9	L5	T5
99.1	L6	T6
100.7	L7	T7

Figure 7.5 An extension block used to summarize payload data.

7.9.2 Custody Transfer

When engineering (or operating within) a challenged network, it is accepted that there will be times when a message will be lost and must be retransmitted. One advantageous way to reduce the penalty for retransmission of data within the network is to reduce the network distance between the point at which the message was lost and the message's intended destination. With BP, the source of a bundle typically accepts responsibility for its delivery at the destination (by requesting an acknowledgment from the destination within a given amount of time). In networks where such acknowledgment cannot be guaranteed (perhaps where a path from source to destination is expected to be unidirectional), delivery is either considered best-effort or the BPA must trust the underlying CL for reliable end-to-end delivery.

The custody transfer mechanism of BP is one in which a waypoint node in a network accepts the responsibility to both forward the accepted bundle and to keep forwarding it until the bundle is received by its destination or until some downstream node accepts custody for the bundle, thereby relieving the interim BPA from the responsibility and transferring it to the accepting BPA. In this way, a bundle source may forget about the bundle once some other node (or nodes) in the network provide a custody acknowledgment. The insight behind this mechanism is that while end-to-end links across a challenged network may be lossy, small clusters of local links representing sections of the path likely have excellent connectivity for some period of time. The concept of custody transfer is illustrated in Figure 7.6.

This figure illustrates the differences in retransmission from a bundle source versus retransmission from a waypoint node that has accepted custody of a bundle. As a function of chained probabilities, the shorter the retransmission path the higher the likelihood of message delivery. In a later chapter we will discuss custody transfer from the point of view of the sending application, and not simply the point of view of the network delivering the data.

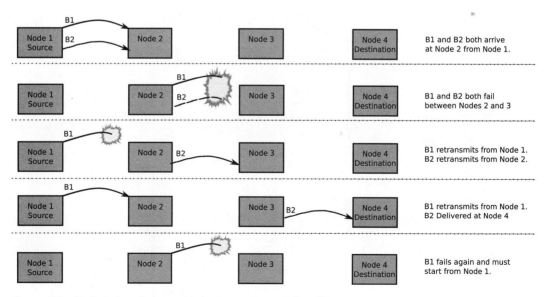

Figure 7.6 Custody transfer concept shortens retransmission distances.

7.9.3 Content Caching

The nominal operational concept for a store-and-forward network is to remove the bundle from storage when the node has successfully transmitted the bundle from itself to the next node. However, there are circumstances where keeping a bundle in persistent storage even after it has been successfully transmitted enables new functionality. Since transmission of the bundle is necessary but insufficient to ensure the bundle successfully makes the hop between BPAs, keeping a bundle until it has been acknowledged by a downstream node (or downstream custodian) is one use case. Another use case is to treat persistent bundle storage as a cache in the event that the storing node decides that other nodes in the network should receive the bundle as well. This is a common circumstance in so-called information-centric networking where data may be preplaced in a network for future consumers. The bundle extension block mechanism, combined with the store-and-forward capabilities of BPAs, provide a significant amount of the functionality necessary for this kind of behavior in a network.

7.10 Special Considerations

While the BP provides several interesting operational concepts, it does not overcome every obstacle encountered in a challenged network. There are several types of problems—some logical and some implementation-specific—that will be encountered by BPAs and particularly BPAs that are not operating over the terrestrial internet. Highlighting these considerations is not meant to dissuade from the use of BP in these scenarios; in fact, BP will often work much better over challenged links than other protocols. However, understanding the limitations of BP will play an important role in developing overall network architectures and designing networking applications.

7.10.1 Storage Management

The benefit of a BPA storing bundles pending transmission (or pending a future requestor as part of a data cache) is offset by the consequence of levying onto nodes the requirement to manage storage. Unlike highly resourced computing nodes, storage is not necessarily inexpensive or plentiful for most practically realizable systems. Resource constrained nodes constantly perform SWAP trade-offs in their design and as a function of what components are operational when. Alternatively, when highly resourced nodes with large amounts of storage are connected to high-rate data feeds, such as those achieved by optical communications (e.g., fiber optic or free-space lasercom links), available storage at the destination end of the link can be filled rapidly. Node designers and network architects must consider the ratio of storage to data rate (for both transmit and receive) in order to determine the likelihood that a node will fill with data as part of regular operation.

As storage on a node fills, the node is faced with a decision: either bundles in the store must be removed (and discarded as lost) or new bundles must be prevented from being stored (and also discarded as lost). In either case, a node will need to review a large amount of policy data to understand how to prioritize its traffic. There are multiple methods for determining which bundles should be lost first due to resource exhaustion, such as:

- Remove lower priority bundles to make room for higher priority bundles;
- Remove older bundles to make way for newer bundles;
- Remove larger bundles to keep a larger number of smaller bundles;
- Remove smaller bundles to store more user data (being contained in the bundle payload rather than the header) versus bundle overhead;
- Remove bundles closer to expiration to make room for longer lived bundles;
- Remove bundles that represent fragments to make room for whole bundles.

Additionally, there are several other considerations as functions of cost, service level agreements, link saturation, and so on. These resource constrained determinations are also separate from the maintenance that must be done as a function of the underlying storage media, such as directory cleanup, garbage collection, and trim support.

7.10.2 Security

The concept of store-and-forward and lack of session information presents fundamental challenges to implementing best practices for security over the terrestrial internet. It is impractical to perform familiar TLS handshakes over a constantly disrupted network. Even if end-to-end connectivity between nodes existed, ready access to security certificate authorities may not exist or the locally reachable CAs may not be up to date. Streaming cipher suites that rely on constant connectivity, in-order delivery, and session state are also unreliable in a challenged network.

Finally, the concept of bundle storage implies that bundles can exist in the network, or on particular nodes for extended periods of time. When storage is measured in hours or days the possibility of a malicious actor intercepting a bundle and decrypting it prior to the original bundle being delivered must be a consideration. Put simply, the longer secured data is resident on a node, or in transmission across the network, the greater the vulnerability to attacks that would be challenging to successfully execute in a non-store-and-forward architecture. This drives additional considerations of cypher-suites, key updates, key management, and time of protection from brute force attacks that are significantly reduced in network designs that can assume continuous connectivity and very low end-to-end transmission times. In store-and-forward models, data across the network must be protected at rest in addition to in motion.

While these are obstacles inherited from the physical and logical limitations of challenged networks and not unique to BP, any communications protocol must find a way to secure its information. In certain cases, this will fall on the applications themselves to supply cipher text as payloads. In other cases, additional BP mechanisms (such as extension blocks) can be used to encode security features within the bundle itself.

7.10.3 Fragmentation

The fact that BP spans such a variety of heterogeneous links increases the chances that a bundle will need to be fragmented. Conceptually, BP fragmentation works in the same way as any other protocol fragmentation: fragments of an original bundle are marked as such and reassembled at the bundle destination over time. Fragments themselves can be further fragmented, and so on. Each fragment is in itself a bundle (with additional annotation noting that fragmentation occurred). The unique challenges with fragmentation in a BP network are twofold: determining an optimal fragment size and deciding how to handle extension blocks.

7.10.4 Optimal Fragment Size

Bundle fragmentation occurs at the BP layer independent of lower-layer CL fragmentation and MTU considerations (which can also be expected to play a role). However, when a BPA has multiple CLAs for multiple CLs, trying to select a fragment size that produces an efficient packing into lower-level MTUs and lower-layer fragmentation protocols can be a challenge. The decision to fragment a bundle at the BP layer rather than allow the bundle to be fragmented by a supporting protocol may be a difficult design choice to make. It will typically fall to network architects and network application developers to encode the policies as a function of exploited knowledge of the network as a whole.

7.10.5 Handling Extension Blocks

When a bundle contains extension blocks that enable the traversal of the bundle through the network, a decision must be made as to what to do with these blocks if the bundle becomes fragmented. If an extension block is meant to be evaluated by waypoint nodes in the network—as in the case of the hop count block or the bundle age block—should it be repeated in every bundle fragment, or should it be included in only a single fragment? If the block is replicated in every fragment and also modified en route (as would be the case with both the bundle age and hop count blocks) how should those blocks be reassembled at the bundle destination? If the block is not replicated in every fragment, how do the other fragments without those extension blocks navigate the challenged network? Figure 7.7 illustrates one potential issue that can be encountered trying to maintain a bundle hop count after a bundle has been fragmented.

In this illustration, bundle B (containing a hop count block showing 6 hops) encounters a BPA that needs to fragment it into three fragments, F1, F2, and F3 which will each take different paths to the bundle destination where they will be reassembled. Whether the hop count block appears in all three fragments, or only one fragment, the semantic meaning of hop count at the destination becomes ambiguous. Further, any attempt by a waypoint BPA to filter bundles based on excessive hop count will be subverted, possibly resulting in the premature expiration of a fragment in the network.

Issues like these with fragmentation are not unique to BP, but BP's reliance on extension mechanisms for state information makes these issues much more difficult

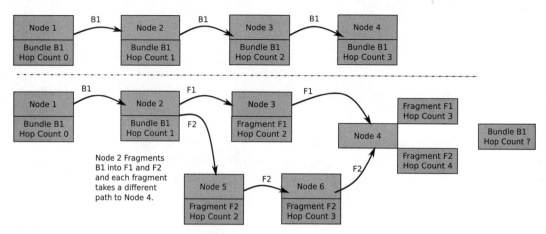

Figure 7.7 The complex interplay between extension blocks and fragmentation.

to solve when they occur. Solutions to the above problem can be engineered into the network, such as not enforcing hop counts on fragments, or defining hop count as the smallest (or largest, or average) number of hop counts taken by any of its fragments. Alternatively, a bundle could be encapsulated in another bundle, with the encapsulating bundle being fragmented to preserve the semantic meanings of these and other types of extensions. These different strategies imply that there is not a single best way to approach these situations.

7.10.6 Additional Processing

Several features of the BP are designed to more compactly represent information under the assumption that the primary burden of nodes in a challenged network is successful transmission between nodes. As seen, blocks in the BP specification are often variable-length and flag fields are used to denote optional fields. While this reduces the amount of work needed for transmission of a bundle, it also requires more processing to occur per bundle at the BPA. When attempting to process high data rates—such as nodes communicative over optical links—the overall node throughput can be restricted by the node's ability to meet this processing demand. The use of non-general-purpose computing mechanisms, such as field-programmable gate arrays (FPGAs) and purpose-built application specific integrated circuits (ASICs), can reduce this processing burden, but their implementation can be significantly complicated by optional and variable length fields. Additionally, while FPGAs and ASICs can provide highly SWAP-efficient implementations of BP and supporting protocols, and can achieve much greater data throughput than GPPs running typical software-based implementations, FPGAs and ASICs are hardware-based solutions requiring additional design, reducing flexibility, and—especially in the case of ASICs—increasing development and implementation time, complexity, and cost of the network appliance.

As the number of bundles increases, nodes must also find efficient ways to rapidly find, retrieve, annotate, and remove stored bundles from the node. If different searches are performed for different reasons (finding bundles for transmission versus identifying bundles to delete, versus querying bundles as part of an information-centric subscription) then nodes may need to efficiently maintain multiple indices of bundles. In some cases, different search algorithms entirely may be appropriate depending on the function, increasing the software (or hardware) complexity of the specific implementation.

7.11 Is BP Enough?

Since BP represents the thin waist of a DTN, it is important to ask the question "is BP enough?" to develop a DTA and to write delay-tolerant applications? Are the features provided by the protocol the very definition of delay tolerance? The answer is that BP is a necessary but insufficient element of the solution to these types of challenging problems.

BP provides several new features and several fewer assumptions than IP networks, but these features require configuration and network services that can react to them, and applications that can populate new fields—such as extension blocks. While the BP provides store-and-forward operation, it does not in itself contain algorithms for forwarding or routing, to understand when a message should be forwarded, and to where. While the BP provides self-annotation mechanisms to augment payloads with information, the BP specification provides very few examples of such annotations and requires additional information relating to network management and security. While the BP provides overlay operation, it does not mandate a particular naming and addressing scheme. Just as the internet is much more than the IP, a DTN requires more than the BP. Such additional definitions, architectural and behavioral models, and protocols have been and continue to be developed (reference the work of the IETF DTNWG) so perhaps it is better to refer to a BP Suite as providing the necessary set of definitions, algorithms, and behaviors for BP to function in challenged delayed and disrupted networks in the same manner that the IPS provides the broader set of features necessary to enable IP to function in nonchallenged networks.

Further, while a DTN provides delay tolerance for messages in a network, an application is not itself a message in the network. Just as BPAs in the network must be designed to implement delayed messages, delay-tolerant applications must be designed to transmit and receive delayed messages—a topic which comprises the bulk of the remaining chapters of this book. Such DTN applications must be aware of and address the challenging conditions present in DTNs (delay, lack of assured connectedness, etc.) in a way that IP applications are not. The BP enables and supports DTN applications and their ability to function effectively in disrupted environments, but again, the BP is not in itself sufficient alone.

Protocols such as the BP can ensure that data is appropriately queued and otherwise handled in the network but cannot apply user data-specific processing semantics because application-layer data is typically opaque to the transport layer. Understanding the emerging properties and limitations of delay-tolerant transport provides the basis for those assumptions that delay-tolerant applications can make, and what capabilities must be handled in the application layer.

7.12 Summary

The BP provides the transport layer by which DTNs can be implemented for data exchange in challenged networking environments. The primary principle motivating the design of this protocol is the use of store-and-forward techniques at nodes in a network. This capability allows nodes to exchange messages even where there is no possibility for end-to-end connectivity and, thus, no synchronized session information.

The PDU of the BP, the bundle is defined as a collection of multiple blocks of information, which allows bundles to include multiple types of information together. For example, a bundle may include both user data and annotative information associated with the data. By allowing various types of data annotations, the BP provides messaging functions that are difficult or impossible to implement using traditional transport layers.

7.13 Problems

7.1 Consider a node which transmits 1,000 PDUs per second providing 40% link utilization over a 1Gbps link. If, for every PDU transmitted (including retransmissions), the PDU has a 35% chance of being retransmitted, what is the ratio of transmitted bits to successfully received bits? What retransmission probability would need to exist to have the ratio be 1.1?

7.2 Assume a network path comprised of four nodes: N0, N1, N2 and N3 with each link between nodes operating at 1Gbps. Each node produces 500 PDUs per second at 20% link utilization to its downstream neighbor with a 25% failure-and-retransmit probability on each link. Node N3 consumes all data produced in the network. Add a new node, N2b which also generates traffic and provides an alternate path between nodes N1 and N3. What is the global ratio of all transmitted bits to all received bits in the original scenario without node N2b? What is the global ratio of all transmitted bits to all received bits in the scenario that adds node N2b? Has the addition of node N2b improved the efficiency of communications in the network?

7.3 Construct a syntactically correct bundle v7 primary block for a message sourced from a NULL node and destined for EID ipn:1.1 with no CRC, a creation time of September 9, 2019 at 9:09pm exactly, a lifetime of 44 years, and no report-to address. You may select any processing flags that you wish. What is the overall size of the encoded primary block?

7.4 Provide an example of using an extension block in a bundle to help a downstream BPA implement a priority scheme. What data would be in this extension block? How would each BPA use this data? Would the BPA need to modify this block before forwarding the bundle?

7.5 Provide an example of using an extension block in a bundle to update downstream BPAs with routing information. What types of information would need to be in the extension block? how would each BPA modify the block as the bundle is forwarded?

7.6 Assume that a bundle has completed three hops when a forwarding BPA decides that the bundle must be fragmented into two fragments, F1, and F2. F1 will take seven hops to reach the bundle destination and F2 will take nine hops to reach the bundle destination. All BPAs in the network have been configured to discard any bundle or fragment that exceeds a hop count of 10. What will the behavior of the network be if: (a) The hop count block is included in both F1 and F2, (b) the hop count block is only included in F1 and not F2, (c) the hop count is included in both F1 and F2, but each fragment is encapsulated in a bundle which itself is given a hop count block starting at zero?

7.7 Assume you are a custody-accepting node receiving a 1Gbps inbound link that receives 1,000 PDUs each second with a link utilization of 50%. You have 500MB of persistent storage. You will accept custody and store 25% of PDUs for later transmission and will need to read from storage approximately 200 PDUs every second from storage for retransmission. Assume for simplicity that the read and write speed of your storage is the same. What does this speed need to be in bits per second to avoid dropping incoming data?

7.8 Provide three additional heuristics that can be used to determine when a bundle should be deleted from persistent storage beyond those already presented.

7.9 Assume that the common networking utility ping has been rewritten to use BP instead of IP. Is ping now able to function in a delayed network? Why or why not?

References

[1] Xu, F., M. Tao, and K. Liu. "Fundamental Tradeoff Between Storage and Latency in Cache-Aided Wireless Interference Networks," *IEEE Transactions on Information Theory*, Vol. 63.11, 2017, pp. 7464–7491.

[2] Kandula, S., et. al., "The Nature of Data Center Traffic: Measurements & Analysis," *Proceedings of the 9th ACM SIGCOMM Conference on Internet Measurement*, ACM, 2009.

[3] Benson, T., A. Akella, and D. A. Maltz, "Network Traffic Characteristics of Data Centers in the Wild," *Proceedings of the 10th ACM SIGCOMM Conference on Internet Measurement*, ACM, 2010.

[4] Hassidim, A., et. al. "Network utilization: The flow view," 2013 *Proceedings IEEE INFOCOM*, IEEE, 2013.

[5] Burleigh, S., "Bundle-in-Bundle Encapsulation, draft-ietf-dtn-bibect-02." Sep. 09, 2019.

[6] Burleigh, S., and K. Fall and E. Birrane, "BP Version 7", Working Draft, IETF, August 2019. Secretariat, http://www.ietf.org/internet-drafts/draft-ietf-dtn-bpbis-14.txt

[7] Castagnoli, G., J. Ganz, and P. Graber. "Optimum cycle redundancy-check codes with 16-bit redundancy," *IEEE Transactions on Communications*, Vol. 38.1, 1990, pp. 111–114.

[8] Castagnoli, G., S. Brauer, and M. Herrmann, "Optimization of Cyclic Redundancy-Check Codes with 24 and 32 Parity Bits," *IEEE Transact. on Communications*, Vol. 41, No. 6, June 1993.

[9] Borrego, C., et. al., "A mobile code bundle extension for application-defined routing in delay and disruption tolerant networking," Computer Networks, Vol. 87, 2015, pp. 59–77.

[10] Birrane, E., and K. McKeever, "BP Security Specification, draft-ietf-dtn-bpsec-11," https://tools.ietf.org/html/draft-ietf-dtn-bpsec-11, Sep. 9, 2019.

[11] Zhang, L., et. al., "Named data networking (ndn) project," *Relatório Técnico NDN-0001*, Xerox Palo Alto Research Center-PARC, Vol. 157, 2010, pp. 158.

[12] Ahlgren, B., et. al., "A survey of information-centric networking," *IEEE Communications Magazine*, Vol. 50.7, 2012, pp. 26–36.

CHAPTER 8
Advanced Networking Architectures

This chapter discusses a variety of advanced networking architectures and concepts that are typically necessitated in challenged networking environments. We have previously discussed how new store-and-forward protocols, such as the BP, enable messaging where networks cover vast spatial distances or experience temporal topological change. This chapter discusses in more detail the nature of the networks within which applications such as BP will find themselves operating. Understanding advanced networking concepts, and in what situations they should be applied, will help network designers determine the architecture and topology of their own network. Further, by defining a precise vocabulary for networking architectures, the role of applications can be better defined. Finally, the use of any type of network architecture is enabled by the features on the nodes that make up the network and this chapter provides the set of features necessary for these types of architectures.

8.1 Networking Architectures

The pragmatic benefit of a hierarchical approach to network protocol design is that individual protocols can be simpler—each protocol relies on the capabilities of the protocols operating beneath it in the hierarchy. Similarly, the set of capabilities provided by network applications can result in simpler user applications. The more functionality supported by the underlying networking application, the less functionality must be written into a user application. When properly designed, network applications should completely insulate a user application from the construction and maintenance of a network. A user application's only concept of network is the one provided to it by its most immediate networking interface. These network interfaces can abstract radically different types of networks, as shown in Figure 8.1.

The ability of network applications to abstract complexity away from user applications becomes particularly important in challenged environments. In such networks, architectures become more complex as nodes operate with less supporting infrastructure and sparse and changing topologies. Apart from the standard networking architectures discussed in previous chapters, there are two types of complex networking architectures that impact the interface between applications and network protocols: overlay networks and federated internetworks. The remainder of this section discusses these networking architectures and some of the network design decisions that must be made when choosing to deploy them.

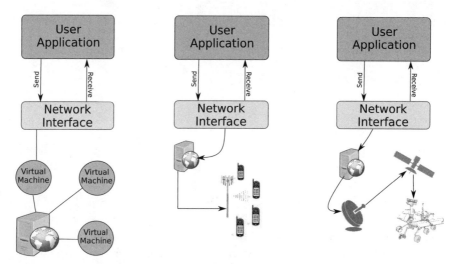

Figure 8.1 Network interfaces completely define the network to user applications.

8.2 A Standard Model for Networking

Any discussion of advanced network architectures needs to happen against the backdrop of a standard network architecture. For the purpose of distinguishing simple from advanced networking concepts, a standard network architecture is one that has the properties of unique identification, point-to-point communication, and topology convergence.

Unique identification means that every physical node in the network can be both consistently identifiable and differentiated from other nodes. To be consistently identifiable means that a node's identity does not change over time as part of the regular operation of the network; if an application is addressing a message to a node and sends a message at a later time to that same address, it may reasonably expect that the message would then go to the same node. Similarly, an application should not need to remember multiple identities for a physical node. Most computers on a home network have this property, with each computer being assigned a unique IP address and a unique media access control (MAC) address.

Point-to-point communications, also called unicast messaging, refers to a paradigm where every message in the network has a single origin (the message source) and a single recipient (the message destination). Network services evaluate these single destinations for operations such as evaluating routes through the network, choosing how to apply security services, and when calculating statistics for network management. It is expected that message sources and destinations are fixed at the time of message creation and that the associations of addresses to physical nodes in the network are stable for the duration of the message's existence in the network.

Topology convergence refers to the ability of a network to reconfigure around one topological change before another topological change occurs. Topology change in a standard network is caused by the addition and removal of nodes in the network. For example, turning on and off a laptop computer represents a change in the topology of a home network. This type of change can be synchronized across the home network in milliseconds.

This standard networking model (see Figure 8.2) requires help from the environment and from the user to exist. The environment must provide sufficient power and communications infrastructure to allow for rapid data exchange without having nodes reset or be constrained in their transmission or computational abilities. It would be difficult, for example, if a home router were run solely on batteries without the ability to plug into a mains power outlet. Also, the overall user data volume must be small enough to not otherwise congest the network and cause significant data degradations. Many home networks are built for at least 1-Gbps data rates but have connections to the terrestrial internet that are significantly slower than this [1] and rarely generate that rate of traffic within their home.

8.3 Overlay Networks

The term overlay network or overlay refers to a networking architecture in which a subset of nodes in a physical network adopt one or more additional virtual identities that form a secondary network that is "overlaid" on the physical network [2, 3]. In this way, each physical node also instantiates zero or more virtual nodes. A physical node will have zero virtual nodes if the overlay network (a subset of the overall network) does not need to source or sink messages at that node. A physical node may have multiple virtual nodes if portions of the overlay network are simulated on that node. In this way, a physical node may support multiple identities: one for its physical network and a virtual identity for each virtual node.

The concept of a virtual network separate from a physical network can be understood by considering a desktop computer running multiple virtual machines. The desktop computer may have a single network interface card (NIC), and thus

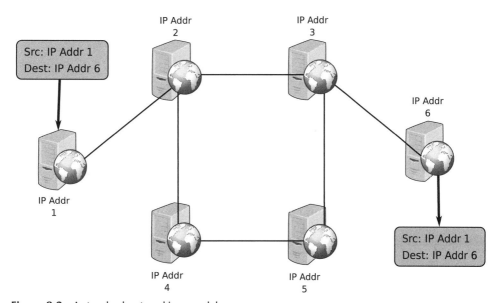

Figure 8.2 A standard networking model.

be a single node on a physical network. However, each virtual machine running on that desktop computer may have its own virtual NIC, with its own virtual address. The physical NIC receives multiple messages over its network connection: traffic destined for the desktop computer itself and traffic destined for each of the virtual machines. In this setup, each virtual machine represents its own virtual node which may or may not be part of an overlay connecting other virtual machines.

Figure 8.3 illustrates an underlying physical network upon which three overlay networks (O1, O2, and O3) have been overlaid. Whereas the physical network consists of 10 nodes, overlay networks O1, O2, and O3 consist of three, five, and seven nodes, respectively. A user application that exists on overlay network O1 does not know about any of the other overlay networks, but it also does not know anything about the physical network supporting O1. The user application only perceives that it is connected to a single network with three nodes. When this application sends a message to another application on some other node in the O1 network, the message is generated by the O1 logical node and transported by the underlying physical node and then provided to the destination O1 virtual node for delivery. These two applications could be multiple hops apart in the physical network but perceive that they are only one hop apart in the overlay. These two applications could both reside on the same physical node running on different virtual machines and perceive that they are multiple hops away in the overlay.

8.3.1 Pass-Through and Encapsulating Interfaces

Overlay networks, by definition, are bound to—but logically distinct from—a physical node. This implies that the virtual nodes comprising the overlay must support some type of interface to the physical nodes on which they reside. There are two ways of designing this interface, pass-through and encapsulation, and the decision

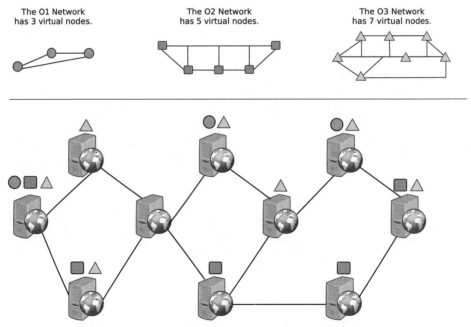

Figure 8.3 Overlay networks are logical subsets of physical networks.

to use one or the other has a significant impact on the functioning of the physical network and any overlays that it supports.

Pass-through interfaces use the same conventions and services in the overlay network and the physical network such that the virtual nodes in the overlay network and the physical nodes in the supporting network can directly exchange information with each other (Figure 8.4). Using this type of interface, the overlay network becomes an extension of the supporting network even though it still requires the supporting network for its message exchange.

One example of a pass-through interface is a physical server on a network hosting multiple virtual servers, where both the physical server and the virtual servers are connected to a corporate network and have distinct IP addresses. In this case, messages from the virtual servers are provided to the physical network without encapsulation or modification and treated as if they were generated from the physical node itself. This means that additional care must be taken to ensure that the virtual nodes are not configured to be in conflict with the physical node or with each other. If two virtual servers were configured with the same IP address this would have the same negative consequences as if two physical nodes on the network had the same IP address.

Encapsulating interfaces prevent any conflicts between the overlay network and the supporting network by never allowing the two networks to intermingle messaging. When using encapsulating interfaces, all messages in an overlay network are completely encapsulated by messages in the supporting network. In this way, the overlay network never sees the naming and addressing of the supporting network and the supporting network never sees the addressing of the overlay network. This clean separation allows the overlay network and the supporting network to use the same address space or different address spaces without issue. Similarly, if the overlay network were to use a completely different addressing scheme (e.g., not using IP addresses at all) it would cause no problem to the supporting network.

In topologies where virtual nodes are required to extend an existing network, pass-through interfaces must be used as this is the mechanism by which virtual nodes and physical nodes can directly interact. A common driver for this type of network extension is when building a modeling and simulation network comprising multiple virtual nodes because the cost of deploying physical nodes is prohibitive. In topologies where the overlay is self-contained, an encapsulation interface should be used, as configurations, security concerns, and management responsibilities can be kept separate between the networks (see Figure 8.5).

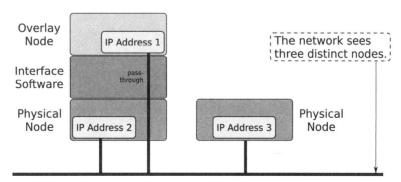

Figure 8.4 Pass-through networks allow overlays to interact with their supporting networks.

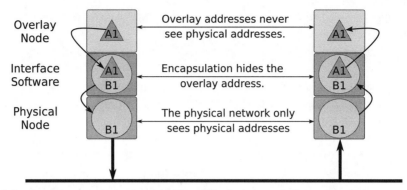

Figure 8.5 Overlay networks are insulated from physical networks by encapsulation.

8.3.2 Differing Network Addressing Schemes

When using encapsulating interfaces between an overlay network and a supporting network, the strong separation of messaging approaches allows for these networks to use completely different naming and addressing schemes. Overlays on the terrestrial internet often use this separation to avoid having to prevent conflicts when using the same naming and addressing schemes, such as the case when IP is deployed on an overlay, as well as a supporting network. In challenged environments, naming and addressing in the overlay may use radically different naming and addressing schemes.

In a typical addressing scenario, messages are sent from a single source to a single destination—a process known as unicast messaging (see Figure 8.6). Most user applications have a concept of the destination to which they are sending their information and most receiving applications (and network protocols) presume that messages have a sole destination, and once delivered, a message can be removed from the network. This is a familiar scenario to anyone who has addressed an e-mail or sent a text message to a single individual. However, unicast messaging is not the only strategy for message addressing, and depending on the topology of the network, may not be an efficient or achievable strategy.

One example of where unicast messaging may not work efficiently is when not every node is uniquely distinguishable. Certain network topologies define convergence nodes as required waypoints along any messaging path that perform common required operations such as security processing, logging, or data exfiltration. To the extent that these nodes all perform the same function in the same way, they may be made indistinguishable from each other to aid in load balancing or to accommodate node mobility. Attempting to route a message to a specific convergence node may be impossible as these nodes may not have unique addresses and specific nodes may come and go in the network as a function of maintenance or node mobility.

Examples of convergence nodes can be found in a variety of networking scenarios. Local area wireless sensor networks typically communicate their data to the terrestrial internet via one or more base stations. Planetary landers communicate their information back to Earth via one or more planetary orbiters. Instrumented ocean buoys communicate their data through one or more LEO satellites.

8.3 Overlay Networks

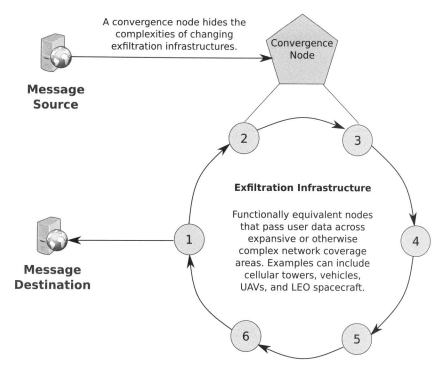

Figure 8.6 Unicast networking assumes nodes are not functionally interchangeable.

In these examples, nodes interact with some provided exfiltration infrastructure whose components are both subject to change and functionally indistinguishable from each other. Attempting to communicate directly to each component (either as a destination or a routing next hop) places an impossible task on the network application: uniquely identify and address elements of a changing and indistinguishable infrastructure. To remedy this source of confusion, unicast networks can deploy a convergence node which accepts unicast traffic from some message source and abstracts from that source the dynamic nature of the underlying network. This architecture works well until data volumes require the addition of multiple convergence nodes. At this point, a message source may no longer understand to which convergence node it should communicate. Common solutions to this problem on the terrestrial internet involve adding convergence nodes for the convergence nodes and building a hierarchy of infrastructure to handle the inefficiencies of unicast messaging.

Separate from undistinguished node addresses, unicast messaging is also problematic in cases where the physical methods of data exchange can naturally communicate with multiple nodes at one time. If a single transmission can be received by multiple nodes within some transmission footprint, then a single message could be received by these nodes for the cost of that single transmission. With unicast messaging, a separate message (and message transmission) would need to be transmitted once for each destination (see Figure 8.7) even if all destinations were in the footprint of each transmission.

One alternative to unicast messaging is to have every node receive and process every message in the network. This strategy is called broadcast messaging and

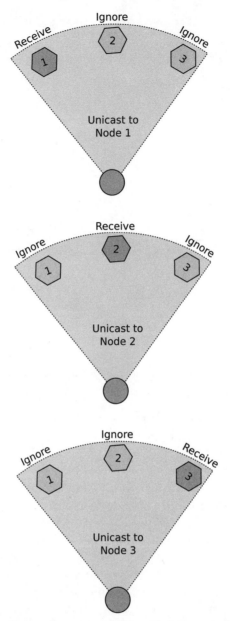

Figure 8.7 Unicast messaging requires a separate transmission per receiver.

broadcast messages are destined for a special address designated as a broadcast address. When a node sees a message addressed to the broadcast address, it processes the message as if it were originally destined for the node. Because broadcast messages are processed by every node that receives them, they are useful for sharing common information quickly. A single broadcast message transmission will communicate a message to all nodes in its transmission footprint, which is a significant gain over a unicast approach which would require one transmission per receiving node. Because broadcast messages do not need a destination node address, they can be used in those special cases where node addresses are not known. For example, when a node is added to a network for the first time, it can send a broadcast message asking for networking information to be sent back to it, with the individual node responses to this broadcast being a series of unicast messages. In this way, a new node can learn the addresses of the nodes within its transmission footprint with very little initial configuration.

The issue with broadcasting as a messaging strategy is that, outside of very specific special situations, broadcasts rapidly consume all networking resources. Broadcasts can reduce the number of transmissions needed to get information to multiple nodes, but every node that receives the broadcast must process the message, and therefore, the overall computational burden in the network is increased. This burden is not significant when a single node makes a broadcast in a special situation, but when all nodes use broadcasting as their strategy the number of messages that every node would need to process increases exponentially. In networks whose connectivity is funneled through convergence nodes (such as hubs, switches, and routers), broadcast traffic can also cause forwarding issues where these devices need to determine whether a broadcast message should be passed through the device, or whether the broadcast came from a part of the network already and passing it through would mean duplicating the broadcast. Generally, when nodes or other pieces of networking equipment fail to properly handle broadcast messages, they may be regenerated and rebroadcasted causing significant congestion in the network. This situation is called a broadcast storm and typically stops most traffic in a network until it passes. This is a particular problem in ad hoc networks [4, 5] where broadcasting is seen as a strategy for overcoming constantly changing network topologies.

A second alternative to unicast messaging is multicast messaging. Similar to broadcasting, which uses a special broadcast address, multicast messages are addressed to a special multicast address. Unlike broadcasting, multiple multicase addresses may be defined in a network, with at least one address per group of nodes that should be receiving information. Similar to overlay networks, each node in a multicast group is configured with multiple addresses: at least one unicast address and one or more multicast addresses. When the node receives a message destined to its unicast address or any one of its multicast addresses, the node will process the message. Otherwise, the node will ignore the message. This strategy is beneficial in that it allows convergence nodes to not need to be distinguished from each other and, thus, hide that complexity from user-layer applications. Because multicast messages have a defined address range, they are differentiable from broadcasts and are less likely to be repeated unintentionally across a network, which reduces the likelihood of multicast storms.

Consider a LEO constellation and a ground station that needs to communicate through the constellation. It is unlikely that the ground station cares which specific satellite it communicates with, as it just wants to get data into and through the constellation and back to its ultimate destination as quickly as possible. If every spacecraft were configured with the same multicast address, then such a ground station would simply forward the message to that multicast address as part of its routing strategy, which would result in delivery to whichever satellite happened to be overhead next or next in the constellation multicast group [6].

When used for messaging, multicast addresses do present other problems for applications. When a multicast message is used to communicate information to multiple users, it becomes difficult to understand whether all intended recipients have received the message and when a multicast message is considered unacknowledged and must be retransmitted. Using multicast addresses to identify the source of a message causes immediate problems when trying to send acknowledgment messages back. For these reasons, multicast addresses are typically only used when the network topology is stable and when links have a very low chance of being disrupted.

Networking overlays provide an architectural approach for abstracting the limitations of unicast messaging without requiring the deployment of multicast or broadcast messages. This is because the virtual addressing provided by the overlay can be deployed independently of the actual addressing of the supporting network. Applications can deploy multicast addressing at the overlay and then implement that addressing as unicast messaging at the physical layer (Figure 8.8), which removes from the application the complexity of managing multiple resources and pushes this complexity to the networking convergence adapters where it can be solved once for all applications operating in the overlay.

This approach is similar to the way in which unicast IP messaging was overlaid onto the early broadcast approach to Ethernet frames. When transmitting over Ethernet, only one message can be active on the wire at a time—transmissions were essentially broadcast to every device on the wire. The illusion of unicast messaging occurred at the IP layer, which was encapsulated within Ethernet frames that had no concept of the IP unicast address. On the wire, a technique called carrier-sense multiple access (CSMA) is used to wait until there is no other traffic on the wire (carrier sense) and then attempt to transmit when things seem quiet [7]. If multiple nodes transmit at the same time and collide, they each wait a random amount of time before trying to transmit again (multiple access). The design of nodes waiting different amounts of time after a collision before retransmitting is what increases the probability of successful message delivery over some time interval. This constant broadcast, collide, back off, and retry that occurs at the Ethernet

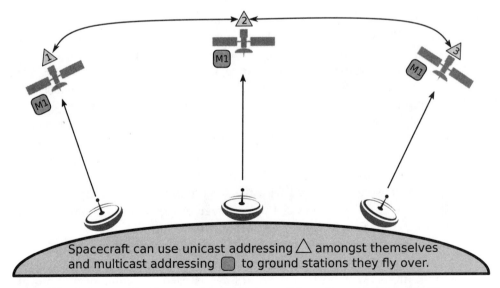

Figure 8.8 Multicast addressing is useful when intermediate nodes are indistinguishable.

layer is completely abstracted from the IP layer, which sees destinations as simple, point-to-point links in the network.

8.4 Partitioned Networks

In cases where networks cannot fully populate a coverage area, individual nodes must be mobile to provide connectivity to different portions of the network at different times. Areas of the network that are connected in the presence of a mobile node and disconnected in the absence of a mobile node are considered a network partition. A sparsely connected network can be imagined as a series of network partitions with one or more mobile nodes used to connect them.

The mobile nodes providing this type of temporal connectivity are termed data mules, reflecting the idea that they transport data between or among otherwise partitioned parts of the network [8]. Data mules may be assigned a multicast address to simplify communications through the network (Figure 8.9). In topologies where network partitions are vastly distributed, no partition might ever see two data mules at once, and subsequently, multiple data mules could be assigned the same unicast address without creating any meaningful collision in the network. Disconnected network partitions may continue to perform local data exchange and collect information (such as by using store-and-forward protocols) waiting for a data mule [9].

The benefit of allowing partitions as part of the network topology is that it provides a lower cost solution for instrumenting a large spatial footprint. This is especially the case when the network has relatively low data volumes. For example, attempting to densely populate the inner solar system such that any spacecraft could talk to Earth, directly or through a relay spacecraft, would be exceedingly expensive. Allowing space-based networks to be partitioned as a function of celestial motion and spacecraft orbits provides a good enough service at a reasonable

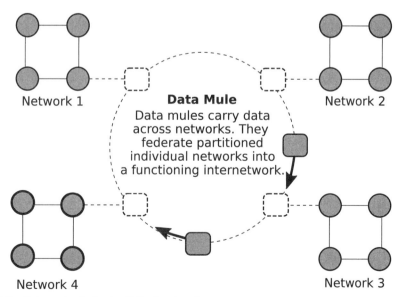

Figure 8.9 Data mules federate distinct networks.

cost. Alternatively, terrestrial internet traffic in a major metropolitan area requires dense node populations to both load-balance traffic and ensure that users are never without connectivity.

8.5 Federated Internetworks

When designing a network architecture that includes multiple network partitions, a question that must be answered is whether these network partitions should be considered part of the same network or whether they can be made their own stand-alone networks. The benefit of having all network partitions be part of the same network is to standardize network services and administrative permissions. This is not always possible, particularly when one network partition may have been built before other parts of the network, or where the vast spatial differences require some partitions to use different technologies than partitions. An alternative to a single partitioned network is to federate multiple stand-alone networks into an internetwork [10].

A federated internetwork is the logical union of two or more stand-alone networks. This is conceptually like an overlay network, with the exception that an overlay exists on a subset of physical nodes within a single network whereas a federated internetwork can span multiple networks each having their own separate implementations, services, and schemes. The concepts of overlay and internetwork are not mutually exclusive, and it is possible to build compatible overlays across an internetwork, particularly when the interfacing method for the overlay is encapsulation. Using the partitioned network model, if a data mule is able to translate or otherwise broker messages among the different technologies implementing the stand-alone networks, then the data mule also becomes the federating node in the internetwork.

In a federated internetwork, federating nodes are the border nodes that can adapt different transport, naming, or other network services from one stand-alone network to another. Consider the flow of science data from a planetary lander to a science operations center. In such a scenario, there are at least four separate networks that must be federated to support this interplanetary data exchange. The first network is onboard the lander and connects the science instrument with the main computer on the lander and with the telecommunications system on the lander. These networks typically have local addressing and very simple routing and access algorithms. Examples of onboard networks include those built around MIL-STD-1553 [11] and Spacewire [12]. The second network exists between the lander and an orbiting relay spacecraft. This network will use a proximity RF link such as the Proximity-1 protocol, whose format and addressing is different from onboard network. The third network exists between the relay spacecraft and ground stations on Earth, where long-haul protocols, such as the telecommand and telemonitoring protocols can be used for high signal propagation delay information exchange. The fourth network is the terrestrial internet which connects the ground station on Earth to a science operation center using familiar networking protocols such as TCP and IP.

Each of these networks is distinct and cannot directly communicate with each other without some federating node performing a translation. The onboard computer acts as a federating node between the local network and the proximity network. The relay orbiter acts as a federating node between the proximity wireless network and the long-haul wireless network. Finally, ground stations act as federating nodes between the long-haul wireless network and the terrestrial internet.

Whenever data must traverse different stand-alone networks, then they are in some way federated, if only for the purpose of the message exchange. Network architects should be able to recognize when this is necessary and incorporate the concept of federating nodes into the topology of the network. A failure to recognize the need to design for internetworking leads to brittle topologies when a poorly federated internetwork needs to incorporate a new network or a new set of message types. One way to reduce the risk associated with this type of internetworking is to consider the types of applications that are necessary to build a network and ensure a separation of responsibilities across those applications.

8.6 Summary

Multiple ways exist in which nodes can be identified and participate in a network. The architectures that represented early computer networks relied on simple notions such as every node having a single address and messaging being defined as between a single source and a single destination. As networks are being extended into more nontraditional areas, this simple architectural approach is unable to scale. Application-layer strategies such as data caching can provide some relief to congestion in certain scenarios, but when nodes must be spatially and temporally separated the fundamental assumptions of the network must be revisited.

Overlay networking abstracts complexity away from user applications but requires a special set of software to handle the interface between the virtual nodes in the overlay and the physical nodes that make up the supporting network. Partitioned networks allow a sparsely populated network to phase its data exchange in synchronization with mobile nodes or other time-based connectivity mechanisms. Federating separate networks allows for data to be exchanged across time and space when a single network is not practical but requires new software to be written to translate between and among the various networks being federated.

Each of these architectures provides new value in certain networking conditions. However, each new architecture also requires software necessary to handle new responsibilities. The hierarchy of applications that must exist to implement, and utilize, these advanced architectures is covered in the Chapter 9.

8.7 Problems

8.1 Provide an example of how layered network protocols allow a simpler application protocol. In your example, explain which functions are implemented by the lower layers and how they would otherwise need to be implement by the application layer.

8.2 Why is an application's only concept of a network the interface provided to it by its most immediate networking interface?

8.3 Name three assumptions inherent in the standard model of networking and provide an example of each, using an at-home network as an example.

8.4 Explain why an overlay network can have more nodes than its underlying supporting network and give an example.

8.5 What are the two types of overlay interface strategies? Provide one benefit and one drawback of each.

8.6 Consider the case of a BPA running on a machine on the terrestrial internet. The BP uses EIDs for addressing and is using a TCP/IP convergence layer. Explain what kind of overlay interface strategy is being used and describe at least three functions that must be performed by the interface software.

8.7 A node just joined a network and is unsure of what other nodes are also on the network. Explain how this node should address a message to the network when it does not know the addresses of any other nodes on the network. How should each node in the network respond to the new node?

8.8 Provide an example of using a multicase addressing scheme to hide addressing complexity from a user application. Explain what happens in your scheme if a node wants to join or leave the multicast group.

8.9 Explain the difference between a partitioned network and a federated internetwork. Provide two benefits for each.

8.10 Consider two separate networks that are federated by a single, immobile federating node and a network architect attempting to add a third network to the internetwork. Discuss three different things that would need to change to incorporate this third network.

References

[1] Sundaresan, S., et. al., "Measuring Home Broadband Performance," *Communications of the ACM*, Vol. 55, No. 11, 2012, pp. 100–109.

[2] Chowdhury, N.M.M. K., and R. Boutaba, "A Survey of Network Virtualization." *Computer Network*, Vol. 54, No. 5, 2010, pp. 862–876.

[3] Jain, R., and S. Paul, "Network Virtualization and Software Defined Networking for Cloud Computing: A Survey," *IEEE Communications Magazine*, Vol. 51, No. 11, 2013, pp. 24–31.

[4] Tonguz, O. K., et. al., "On the Broadcast Storm Problem in Ad Hoc Wireless Networks," *2006 3rd International Conference on Broadband Communications*, Networks and Systems, IEEE, 2006.

[5] Wisitpongphan, N., et. al., "Broadcast Storm Mitigation Techniques in Vehicular Ad Hoc Networks," *IEEE Wireless Communications*, Vol. 14, No. 6, 2007, pp. 84–94.

[6] Ekici, E., I. F. Akyildiz, and M. D. Bender, "A Multicast Routing Algorithm for LEO Satellite IP Networks," *IEEE/ACM Transactions on Networking (TON)*, Vol. 10, No 2, 2002, pp. 183–192.

[7] Gelenbe, E., and I. Mitrani, "Control Policies in CSMA Local Area Networks: Ethernet Controls," *ACM*, Vol. 11, No. 4, 1982.

[8] Shah, R. C., et. al. "Data Mules: Modeling and Analysis of a Three-Tier Architecture for Sparse Sensor Networks," *Ad Hoc Networks*, Vol. 1, Nos. 2-3, 2003, pp. 215–233.

[9]   Zhao, W., M. Ammar, and E. Zegura, "Multicasting in Delay Tolerant Networks: Semantic Models and Routing Algorithms," *Proceedings of the 2005 ACM SIGCOMM Workshop on Delay-Tolerant Networking*, ACM, 2005.

[10] Birrane III, E. J., *Virtual Circuit Provisioning in Challenged Sensor Internetworks with Application to the Solar System Internet*, University of Maryland, Baltimore County, 2014.

[11] Bracknell, D., "Introduction to the Mil-Std-1553B Serial Multiplex Data Bus," *Microprocessors and Microsystems*, Vol. 12, No. 1, 1988, pp. 3–12.

[12] Parkes, S. M., "SpaceWire–The Standard," DASIA 99–Data Systems in Aerospace, 1999, pp. 111–116.

CHAPTER 9
Application Services and Design Patterns

This chapter discusses the types of application services that exist in a variety of advanced networking architectures and how patterns can be created for their design and implementation. Some types of service provide the basic transport functions of a network whereas others provide the end user data production and consumption for which networks exist. In overlay networks, certain types of services are developed to maintain the function of the overlay and are treated as a user service on the underlying physical network and as a network service for the users of the overlay. Internetworks may define types of services to handle physical and logical differences between and among their constituent networks. An understanding of the types of services that exist in a network and across an internetwork is necessary to scoping and building both modular architectures and reusable applications. Attempting to implement multiple types of service in a single application makes those applications (and the networks they enable) brittle in the presence of faults and expensive to modify when new services are needed.

9.1 A Multitiered Application Service Hierarchy

Networking applications operate within the context of the networks to which they are deployed. Their structure, capabilities, and limitations are informed by the assumptions they can make on the networks and the constraints imposed on them by their networks. However, these same applications are also the ones implementing the responsibilities of the networks—their data planes, control planes, and data producers and consumers.

Certain networked applications implement the protocols that keep a network operating, such as those that distribute routing and naming information. Other applications provide data services, such as ensuring in-order, nonduplicated data delivery. Other applications perform more direct services for users: generating and consuming user data, such as delivering user text messages or social media updates. End-to-end data exchange occurs not as the result of a single user application, but as a result of the coordinated effort of multiple networked applications.

Just as networking protocols can be segmented into different layers, applications can be logically differentiated based on the types of services they provide. Simple network architectures define three types of services: transport, node, and endpoint services. In this model, there is a many-to-many mapping of applications to services (see Figure 9.1). In some cases, a single application may implement

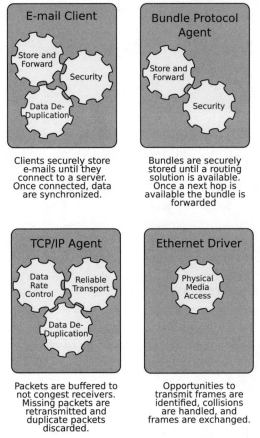

Figure 9.1 Applications implement one or more services.

multiple services in the context of a network. In other cases, multiple applications might cooperate to provide a single networking service. Throughout this chapter, the term application refers to a physical software element in a network whereas the term service refers to a logical feature implemented by one or more applications.

Simple networking architectures rarely provide the features and characteristics necessary to implement data exchange in complex environments. Advanced networking architectures such as overlays, federated internetworks, and federated overlays allow user applications to maintain one or more logical abstractions of a network that can be very different from the actual supporting physical networks. The brokers and adapters that provide these abstractions to user applications are, themselves, services in the network. The traditional decomposition of network services must be expanded to draw more meaningful distinctions between and among the types of applications operating in, and needed for, a functional network.

The three-tier service model can be expanded into a five-tier model (see Table 9.1) by adding two additional tiers of services to cover overlay behavior and internetworking behavior. These service tiers represent a logical separation of responsibilities, and in a perfectly modular network implementation, network applications would provide only services in a single tier. More practically, it is likely that an application implements services from multiple tiers, either as a function of conserving computing resources or using legacy or already-existing components.

9.1 A Multitiered Application Service Hierarchy

Table 9.1 Service Types

Type	Service	Responsibilities
I	Transport	Format messages for transmission over a physical medium with the behaviors and features enabled for various network transport protocols.
II	Networking	Configure and use transport services on a node-by-node basis, including responsibilities such as naming and addressing, routing, and quality of service enforcement.
III	Federating	Broker core services between different physical networks.
IV	Overlay	Convert from logical addressing schemes to physical network addressing schemes.
V	Endpoint	Serve as the sources and sinks of user data in the network.

Regardless of the mapping between application and service, a structured definition of services provides network architects and software developers the terminology and concepts necessary to assess the modularity of the network and how that modularity might be improved.

The service tiers organize into a loose hierarchy—lower tiered services tend to implement functions that are useful to higher tiered services. The scope of services that deal with physical networks (Tier I) to services that deal with abstractions (Tier V) does imply a hierarchical set of dependencies (see Figure 9.2). For example, user endpoint services will exchange data using core network services which, themselves, use transport services. This hierarchy does not imply a strict layering. In cases where there is no concept of an overlay or a federated internetwork those tiers

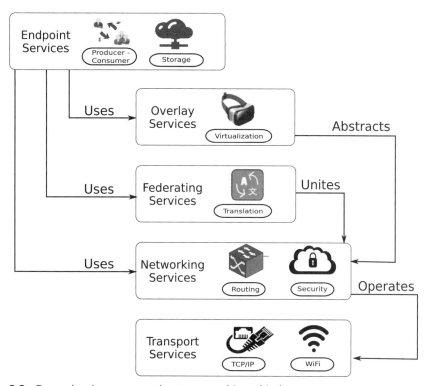

Figure 9.2 Dependencies across services types are hierarchical.

of services will not exist, and Tier V services will rely directly on the capabilities provided by Tier III services.

9.1.1 Transport Services

All nodes in a network, regardless of the characteristics of the network, transport user information. Transport services are fundamental to the concept of networking and must exist for a network to function. Services in this tier implement the protocol agents used to format, process, and generally send/receive information. The association of transport services and protocols is strong enough as to be the defining characteristic of the tier; a service can be said to have Tier-I responsibilities if, and only if, the service implements a transport-layer networking protocol.

The specific responsibilities of transport services are those associated with protocol implementation. This includes creating the messages defined by transport protocols and processing them in accordance with required transmit and receive behaviors. In this context, messaging implies the appropriate framing of user information (payloads) for network transmission. At this tier of the service hierarchy, user data is opaque and not involved in any decision-making regarding protocol implementation or compliance.

Services in this area can either be independent or have dependencies as a function of network protocol layering. In cases where one protocol relies on an underlying protocol for some network services, then the transport services implementing these protocols would have dependencies on other services in this tier and possibly be implemented as a single application. Consider a standard terrestrial internet use of a TCP/IP/Ethernet stack (see Figure 9.3). In this case it is expected that multiple transport services would exist; one implementing the TCP protocol, another implementing the IP protocol, and a third implementing Ethernet protocols. These services may exist in different parts of the node, with the TCP and IP stack being implementing in the underlying operating system and the Ethernet protocols being implemented within the drivers of a NIC. Despite how they might be implemented,

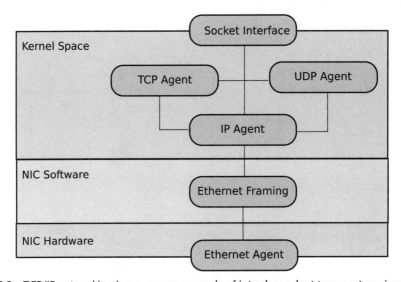

Figure 9.3 TCP/IP networking is a common example of interdependent transport services.

they do have dependencies on each other to be able to be called in the proper order with the proper inputs for both sending and receiving various messages (TCP, IP, and Ethernet messages).

9.1.2 Core Networking Services

Core network services are those services running on nodes in a network in support of common network functions. These are the services that exist between the transport layers and the user applications in a typical network configuration. They advise how, when, and where user data should be sent. These services include the control plane for the network—the set of services that command and control the overall functioning of the network transport.

These services depend on the type and topology of the network but have the overall responsibility of determining which transport services should be used, and with what configuration, for user data exchange (see Figure 9.4). This includes standard network functions such as applying security policies and implementing routing and forwarding algorithms. This can also extend to enforcing service agreements and implementing quality of service mechanisms. This area also includes the services used to generate, maintain, and report on the health and status of these services.

For example, a common core network service is forwarding. Given a destination, forwarding services generate either an end-to-end path to that destination or (at a minimum) the next hop to take on the way to that destination. A forwarding service may implement algorithms that operate on network topological data to understand what paths are available. Separate services may also exist to measure

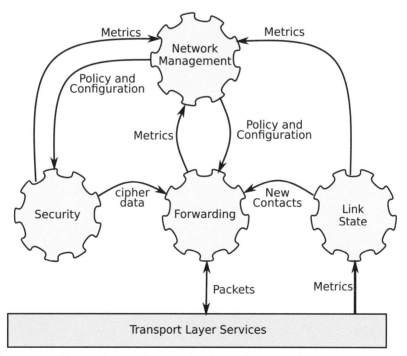

Figure 9.4 Networking services configure each other and transport layers.

the signal quality of links between nodes, to predict when the network topology might change, and to communicate these changes to other nodes to inform their own forwarding services.

9.1.3 Federating Services

Federating services broker information between different physical networks that have been federated into an internetwork. The concept of federation in this sense refers to the need to broker differences between and among the various physical networks. In cases where different physical networks all support the same features and configurations, their internetwork requires no special federating services. Tier III services only exist in federated internetworks and only at the nodes forming the brokering intersection between physical networks. They are necessary because different physical networks may user different transport protocols (and thus different transport services) as well as implement different core network services.

Federating services must account for differences that exist in the messaging formats, behaviors, policies, and configurations among the various physical networks comprising the federated internetwork. In some cases, this may be accomplished through a mechanical translation of common information from one format to another, such as removing message headers for one protocol and replacing them with headers for a different protocol. In other cases, a more complex synthesis of information must be performed when different networks support features or configurations that are not one-to-one compatible with the features and configurations of another network. In addition to mapping between physical network differences, services in this tier must also determine the appropriate physical networks to use for both user messages and traffic generated by core network services. They must determine what kind of networks are being spanned by the node and determine the correct mappings through these networks.

Due to the vast distances covered, space exploration provides many practical examples of federated internetworks. Consider the end-to-end data flow of an image from a camera on a rover on the surface of Mars through an orbiting communications relay spacecraft and back to a science operations center connected to the terrestrial internet (see Figure 9.5). The image, in this case, must traverse at least four networks that have different physical characteristics, different transport layers, and different core network services.

The first physical network exists on the rover itself and connects the camera to the central computer and the telecommunications system using some onboard protocols such as MIL-STD-1553. The second network exists between the rover and its various communications relay orbiters using a different transport such as the CCSDS Proximity 1 protocol. The third network exists between the communications relay orbiter and ground stations on the Earth, using a different transport designed for long-haul communications, such as the CCSDS Telemonitoring (TM) protocol. Once on Earth, the image can be transported from a ground station to a science operations center using terrestrial internet protocols, such as TCP/IP. From the point of view of the image data, it traverses a single internetwork, but the data exchange only works because of the federating services acting as translators and brokers at each of the gateways between the four physical networks.

9.1 A Multitiered Application Service Hierarchy

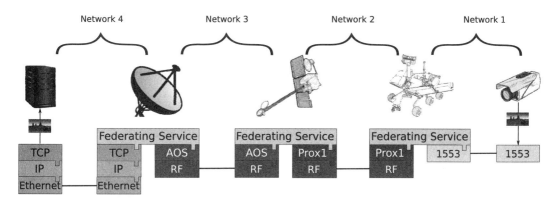

Figure 9.5 An end-to-end federated internetwork between Earth and Mars.

9.1.4 Overlay Services

While federating services correct for differences between physical networks, overlay services correct for logical differences within a single network. Overlay services implement virtual networking services that are independent from, and overlaid on top of, the networking services of the supporting physical network. The value of these services is that they provide a simplified, logical view of the overlay network to endpoint applications. This simplified view can significantly differ from the actual physical network, including different sizing and complexity of naming and addressing schemes, smaller or larger numbers of logical nodes for message exchange, and a dramatic impact (increase or decrease) to the overall amount of network messaging as a result.

Overlay services map logical network services to physical network services. This mapping does not need to be one-to-one; a single overlay service can be implemented using multiple physical layer services, and multiple overlay services can be implemented through a single physical network service. The concept of mapping a single overlay service to multiple underlying physical network services happens when the overlay network provides a feature that is otherwise not present in the underlying network.

An example of an overlay network is the use of the BP over the terrestrial internet. The BP uses its own naming and addressing scheme (EIDs), different routing algorithms, different security protocols (BPSec), and different network management protocols. The BP concept of a CLA implements overlay services. CLAs translate naming and addressing associated with BP to the naming and addressing of the underlying transport protocol. CLAs may also provide inputs to the BP layer for routing, management, and security information.

9.1.5 Endpoint Services

Endpoint services exist at the endpoints of a messaging exchange. This terminology can be taken literally, in that there can be a single application at a message source and a single application at a message destination. These services also apply when an endpoint is less literal, such as when storing data in a data cache or when messaging to multicast destinations. In this context, endpoint means any node at which a network feels that it has delivered user information out of the network space and into user space.

Because endpoint services represent the points at which data leave the network, they are only responsible for the creation and consumption of user data, and never the implementation, configuration, or management of other tiers of services. They can only comprehend the existence of the network through whatever network interface is provided to them by other tiers. The goal for all other services in the hierarchy is to enable developers to write simple endpoint applications even in the presence of very complex network architectures and topologies.

9.2 Application Design Patterns

Any functioning network is customized by the specifics of its deployment and operational maintenance. While individual networks may have their own unique characteristics, they often derive from a finite set of baseline architectures and exist to solve one or more common types of problems. To the extent that these baselines share common characteristics, they can be viewed by network architects and software engineers as architectural patterns to be followed when constructing networks or individual network services. An example of an architectural pattern is an overlay network—while individual overlays may be distinct, they all share common properties and best practices and help both architects and engineers to reason about commonly encountered technical issues.

Just as architectural patterns help with the construction of architectures, an application design pattern (ADP) helps with the design of applications. An ADP is a portion of a software architecture that solves—partially or completely—a problem likely to be encountered in the implementation of a service in the context of an application. In the context of a delay-tolerant networking application, an ADP addresses the types of problems that may be encountered in the construction of the application in the context of a DTN operating in a challenged networking environment.

Application designs for features such as content caching, rule-based autonomics, and store-and-forward operation each are defined by a core set of principles and responsibilities that can be patterned and used as the basis for a unique implementation. Application design patterns can have a greater impact than network architecture patterns because networks are implemented by applications. Any application's notion of, and interface into, a network is completely encapsulated by the applications with which it communicates and the services those applications provide.

For example, an application providing end-user services may have no understanding of whether it is running over an overlay, participating in an internetwork, or existing within a disconnected partition. The services present by the set of applications in the network determine the practical management and utilization of the physical links in the network. The data volumes produced by applications describe the rates of physical links that must be present. The ability to store and preplace information determines the number of links that must be present. The quality of service associated with data determines loading and redundancy.

It is not possible for a network architect to assess the proper architectural design of a network without understanding the services that will implement the network. Because network architecture is often determined early in a network development effort, the exact form of these services, and how they might be coalesced into various applications, may not be known. Having a compendium of common networking services provides an effective way to reason about the types of capabilities that might be present in the network with enough fidelity to inform the architecture of a specific network.

9.2.1 The History and Concept of Design Patterns

The concept of common designs for common problems (modularity and reuse) has been part of the discipline of software engineering since its inception. These principles became more prevalent and standardized with the popularity of object-oriented programming languages and the general concept of object-oriented analysis and design (OOAD) principles. A brief review of the strengths of an OOAD approach to application design provides a useful context for the discussion and formatting of ADPs.

One of the outputs of a traditional OOAD effort is a decomposition of responsibilities that must be performed by a piece of software. By understanding the full set of such responsibilities, a software developer can determine where there is overlap across the responsibility set. When multiple responsibilities have significant enough overlap, object-oriented languages provide the mechanisms for representing them in a less redundant manner, such as using interfaces (popular in the Java programming language) and abstract base classes (popular in the C++ programming language).

An abstract base class (ABC) approximates an ADP because it represents a common design meant to solve a common problem (Figure 9.6). The C++ term base class refers to a language construct that defines a common (base) set of interfaces, algorithms, and/or implementations shared by multiple portions of a software application (or multiple software applications). The term abstract refers to the fact that the ABC does not have enough information, on its own, to provide

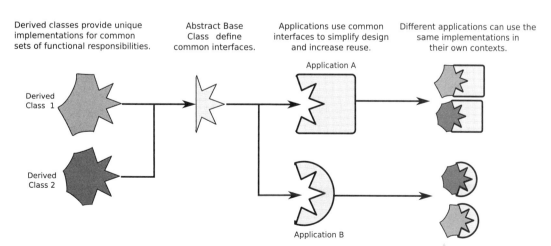

Figure 9.6 An abstract base class provides a default interface and design.

a stand-alone implementation—it is a necessary but insufficient set of the overall software needed to implement an application. Object-oriented programming languages such as C++ also allow nonabstract base classes providing a default implementation that can be instantiated on its own. However, the abstract base classes most closely fulfill the functions typically attributed to design patterns.

A familiar example used when discussion OOAD principles is the representation of, and operations on, geometric shapes. In this example, an ABC can define the set of data representing common identifying information for shapes, the implementation of functions common to all shapes, and the interface used to implement those operations that might be specific to individual shapes. Given this information, a developer who wishes to implement a new geometric shape inherits the design work that went into defining geometric shapes and provides only the delta information associated with characteristics specific to the shape.

There are multiple benefits of using ABCs (or their equivalents) in a modular software design. First, an ABC reduces the overall amount of software that needs to be written by an application. Since an ABC represents the set of common software that is shared by multiple parts of an application, consolidating that into a single area removes the need for the same software to be written multiple times. Second, an ABC provides a standard interface for how these common software components should be used within their application. This guarantees that those who conform to the interfaces provided by the ABC will implement the functions in a standard way and therefore those developers do not need to rederive or redesign those interfaces. Finally, an ABC provides the ability for software reuse across applications. If software is written to the interfaces provided by an ABC, then any other software that conforms to that interface can be used interchangeably.

The concept of design patterns became a natural extension of this ABC concept. While an ABC provides a prefabricated design and implementation paradigm for a specific inheriting class, a design pattern could lead to the design and implementation of ABCs and other software across multiple applications existing across multiple technology domains. The seminal work in defining ADPs and their beneficial impact is Design Patterns: Elements of Reusable Object-Oriented Software [1]. First published in 1994, this book provided formulations of design patterns as a useful universalization of object-oriented design principles. Through its more than 39 reprintings, this book and the work of its authors significantly altered the way in which reuse and maintenance of designs (not just implementations) changed the discipline of software engineering. In 2005, the authors of this book received the ACM Programming Languages Achievement Award for their work on design patterns.

9.2.2 The Value of Patterns in Emerging Application Domains

The promise of a design pattern is that it offers a reusable solution to a commonly encountered problem. When building new applications, software engineers cognizant of relevant patterns in their field can determine which patterns apply to their situations. This can reduce the cost, schedule, and risk associated with developing new software or adapting existing software into new environments. The successful application of a design pattern also indicates that the problem domain was sufficiently understood by the architect and engineering teams.

One way to determine whether a design pattern is a good solution to a problem is to observe examples of where the pattern has been used successfully. This is a reasonable task in mature application domains with multiple application implementations and a long history of operation, maintenance, and lessons learned. When attempting to assess the quality of a patterned solution in an emerging domain, the overall data available for analysis may either by much smaller or much harder to translate to the emerging domain. An emerging domain does not have a significant history of operational deployment to provide a basis for assessing the benefit/value of the proposed patterns.

Some discussion on the limitation of patterns and the uniqueness of the emerging domain is necessary to determine the utility of a design pattern approach. While good design patterns are known good solutions to common problems, they may be terrible solutions to other types of problems, and their indiscriminate application can cause more problems than those being solved. Understanding when and how to apply a design pattern is as much a key to software success as the pattern itself. Some reasons why a design pattern may not be applicable to a software problem include problems with the pattern itself, problems with the software being patterned, or problems translating the pattern in a new context.

There are two common reasons why a design pattern may represent a poor solution to a common problem. First, the pattern may be addressing limitations in a programming language, hardware implementation, or other constraint rather than something inherently part of the software design. If the pattern is really providing a work-around, then when the underlying issue is corrected, such as when hardware and language evolves, the pattern may lose most of all of its utility. In some cases, as the work-around becomes less necessary such a pattern can introduce inefficiency and other problems. Second, a design pattern may not provide a good solution to the problem at all, regardless of other issues. When a design pattern, properly applied, does not provide the desired results it is termed an anti-pattern. Anti-patterns are common solutions to common problems that are inefficient, ineffective, or create more problems than they solve [2] (see Table 9.2).

Applying a design pattern to existing software may cause problems based on the nature of the software rather than the nature of the design pattern. Design

Table 9.2 Sample Design Anti-Patterns

Anti-Pattern	Description
Big ball of mud [3]	Systems that accumulate features over time with no discernable internal architecture. Also called spaghetti code. May indicate the presence of brittle code, lack of modularity, and insufficient requirements analysis and design.
Abstraction inversion*	Hiding necessary functionality behind an abstraction and making those on the other side of the abstraction reimplement the functionality on their own. May indicate a poor understanding of data and control flows across interfaces.
Interface bloat	Making an interface so powerful or customizable that objects cannot conform to the interface. May indicate insufficient requirements analysis or the need for multiple interfaces.
Stovepipe/ not-invented-here	Building custom features and functions rather than incorporating those functions from others. May indicate nonmodule software and increased risks from unnecessary development.

*http://wiki.c2.com/?AbstractionInversion.

patterns are, by definition, part of the design of a software application. In cases where the software design has already been set, such as when refactoring, adding new features, or fixing issues, the impact of applying patterns must be considered. If an existing software design does not provide a natural way to incorporate a pattern, then either the overall design must be reconsidered, or the pattern must be adapted or abandoned. Assessing whether software should be refactored and incorporate a new pattern is a function of good engineering judgment and the resources available to make and test changes.

One memorable rubric for determining whether and when it is time to refactor software is when the software has adopted a bad smell. A code smell, as defined by Kent Beck and Martin Fowler [4], is a characteristic of software that might indicate a problem with the design or the implementation. Smells imply more subtle concerns than those apparent from anti-pattern. Anti-patterns are known problems in software design whereas smells are hints that something might be going bad. Code smells can be introduced from the start of a bad design or as part of inefficient maintenance [5]. While code smells focus on issues in implementation, architectures in software can also smell [6–8] (see Table 9.3).

Even when a design pattern is appropriate and implementable in an existing software base (or as part of a new software development effort), care needs to be taken. By its nature, a design pattern is a generalized solution, which means that the proper implementation of the pattern may need to consider multiple operational scenarios. Software designers must consider which of these scenarios are credible for their application, paying attention to the type of application being written. Design patterns should be customized where practical and when software implementers are knowledgeable regarding the value of the pattern and the assumptions that can be made on how an application interacts with other applications in its ecosystem.

9.3 The Design Pattern Documentation Format

While each of the design patterns provided in this book are different, they are documented using the common format provided in this section. This design pattern template consists of seven sections that address the description of the pattern, how it should be implemented, and how to balance when it should be used. Network architects should be conversant with these patterns when assessing the applications that help implement networking features. Application developers should use these

Table 9.3 Sample Architectural Smells

Architecture Smell	Description
Design by committee [9]	A committee, presumably with multiple interests, produces an overly complex design that seeks to solve too many problems.
Jumble [9]	Poor separation between horizontal and vertical design elements leading to high coupling among components and a brittle design.
Too many layers [10]	Multiple indirections in software require inefficient layers that exist only to call functions in some other software layer.
Feature concentration [8]	A single architectural element attempts to solve multiple architectural concerns.

patterns where applicable to build modular applications. End users should be aware of these patterns when developing operational concepts and workflows to ensure that they are using applications correctly.

The design pattern format in Figure 9.7 breaks down pattern documentation into 7 sections. These sections are used as the outline for Chapters 10–15 that describe the design patterns presented in this book. The pattern context briefly describes the networking environment which presents a common problem and why there is likely a common solution to that problem. This may include a discussion of how certain algorithms in a network are developed, how challenged networking environments constrain communications, or some other insight into the methods and behaviors of DTNs. For end users, the context alone may provide enough information to adjust thinking on operational constructs and how to federate multiple applications to accomplish common goals. Where applicable, this section will also include an analogy when doing so will help the reader reason about the pattern by matching it to a familiar example.

The "problem being solved" section helps architects and engineers determine whether the pattern should or should not be considered. As previously discussed, force-fitting patterns to problems, architectures, or code bases can create more problems than the pattern solves. Having a good understanding of whether a pattern applies in a given situation is the most significant way to avoid the pitfalls of a patterned approach to network and software design.

The "pattern overview" section provides an overview of the roles and flows that exist when implementing the pattern. Pattern roles refer to logical functions that must be performed by the pattern and may not translate one-to-one to any physical design component in software. A logical enumeration of control and data flows allows architects and engineers to adapt these functions to a physical design in a way that is efficient in the context of a particular network or application. Providing more normative guidance than necessary can lead to over-specifying the

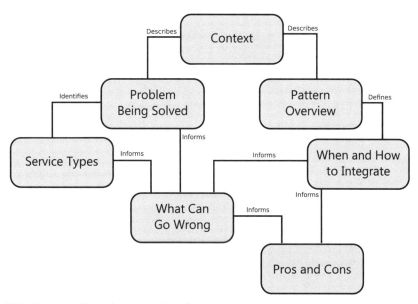

Figure 9.7 Design pattern documentation format.

problem and make the translation of a pattern into software more complex, less efficient, and more brittle than it would otherwise need to be.

The "service types" section addresses the types of services that would likely benefit from using the pattern. As previously mentioned, complex networking architectures use different types of services to separate roles and responsibilities. When a single application attempts to fulfill too many such services, the application becomes tightly coupled to architecture and assumptions. This makes it difficult for the application to adapt as the architecture adapts. Network architects can use this information to decide whether and where services should exist within a network. Software developers can use this information to determine how and how many services their applications should implement and whether their application needs become assumptions or constraints on other applications in the network.

The "when and how to integrate" section provides guidance on how to evaluate whether an existing software codebase or networking architecture should or should not be refactored to accept this pattern. Even in cases where the pattern is useful, attempting to implement it without an understanding of how the pattern integrates with an existing system can cause other, larger problems in either a specific implementation or in the overall performance of the network.

The "what can go wrong" section provides examples of specific issues that can arise in the implementation or use of applications that conform to this pattern. By listing specific examples of issues that can be experienced, architects and engineers can more equitably reason about whether the pattern will succeed in achieving specific goals within an appropriate risk posture.

The "pros and cons" section summarizes the overall disposition of the pattern in a way that helps architects and engineers quickly determine whether a pattern should or should not be applied. This section finishes the template and summarizes the impact of the pattern as discussed in greater detail in the preceding sections.

9.4 Summary

Complex networking architectures—to include those used to build delay/disruption tolerant networks—are implemented by their supporting applications. Part of the responsibility of the network architect is identifying required logical roles and responsibilities and mapping them to the multiple applications that implement the network. Part of the responsibility of software engineers is to build applications in a modular and maintainable and recombinable way to allow for future interfaces and architectures.

One way to help enforce these goals is to define a series of application types to formally differentiate the ways in which different applications enable different features in a network. This chapter identified five different types of networking services that work with each other to implement advanced networking capabilities and the applications that need them. Network applications should be designed to perform a small set of highly cohesive activities in a loosely coupled, modular way. By not overloading design decisions in a single application, networks can be evolved to incorporate new types of links, new types of data flows, and new environments (such as those that add delay and disruption). The ability to mix and match small applications provides more flexibility for architects and end-users whereas monolithic architectures and multifeatured applications make networks brittle and hard to change.

Flexible networks consisting of many well-scoped applications is good architectural practice, but not a guarantee that future maintenance will be simple or that the network will always work in any of a variety of increasingly challenging environments. Regardless of the specific architecture, certain environments manifest problems that recur repeatedly. Discussing common solutions to those problems allows architects and engineers to leverage past experiences and lessons learned to solve issues in as efficient a way as possible. Delay-tolerant design patterns capture proposed solutions to common problems found when implementing DTNs and the applications that must exist within them. These patterns specify a logical set of roles and responsibilities for accomplishing tasks and discussing how they should be implemented across the different application types.

Together, a knowledge of common application types and patterned solutions to common problems create a vocabulary for discussing the precise ways in which DTAs are unique.

9.5 Problems

9.1 For each of the five application responsibilities, list an example of a protocol implementation or other application that provides those logical responsibilities.

9.2 Provide an example of a single protocol implementation or other application that performs more than one type of application responsibility. Discuss whether performing multiple responsibilities makes the application more efficient. Discuss whether performing multiple responsibilities makes the application less adaptable.

9.3 Assume an application performs both transport and overlay responsibilities. Provide an example of why such a design choice could help a single network implementation. Provide an example of why such a design choice would make the application more difficult to port to a different network or networking architecture.

9.4 Explain why the application responsibility tiers should be considered loosely hierarchical. Provide one example where applications might not want to follow a strict hierarchy.

9.5 List three common "core network services" that must be present in every network. Explain why these are core network responsibilities and not part of any other responsibility tier.

9.6 Describe the difference between a federating service and an overlay service. Provide an example where both services can be performed by the same application. Provide an example where they cannot.

9.7 Describe three possible negative consequences of applying a design pattern to an existing software codebase.

9.8 Independently research two design patterns that are not presented in this textbook and describe which responsibility tier they could be applied to.

9.9 Independently research one code smell and describe a situation in which that smell indicates a need to refactor software. Describe a situation in which that smell does not indicate a need to refactor software.

9.10 Discuss how the concept of code and architecture smells can be used in the decision to apply a design pattern.

9.11 Describe two differences between logical roles and responsibilities in an architecture and structural elements of a detailed design.

9.12 Provide and defend three reasons why a design pattern should list logical roles and responsibilities instead of structural design elements.

References

[1] Gamma, E., *Design Patterns: Elements of Reusable Object-Oriented Software*, Pearson Education India, 1995.

[2] Webster, B. F., *Pitfalls of Object-Oriented Development*, M&T Books, 1995.

[3] Foote, B., and J. Yoder, "Big Ball Of Mud," *Pattern Languages of Program Design*, Vol. 4, 1997, pp. 654–692.

[4] Fowler, M., *Refactoring: Improving the Design of Existing Code*, Addison-Wesley Professional, 2018.

[5] Tufano, M., et. al., "When and Why Your Code Starts to Smell Bad," *2015 IEEE/ACM 37th IEEE International Conference on Software Engineering (ICSE)*, Vol. 1, pp. 403–414.

[6] Garcia, J., D. Popescu, G. Edwards, and N. Medvidovic, "Identifying Architectural Bad Smells," *2009 13th European Conference on Software Maintenance and Reengineering*, Kaiserslautern, 2009, pp. 255–258.

[7] Garcia, J., et. al., "Toward a Catalogue of Architectural Bad Smells," *International Conference on the Quality of Software Architectures*, Berlin: Springer, 2009.

[8] de Andrade, H. S., E. Almeida, and I. Crnkovic. "Architectural Bad Smells in Software Product Lines: An Exploratory Study," *Proceedings of the WICSA 2014 Companion Volume*, ACM, 2014.

[9] Brown, W. H., et. al., *AntiPatterns: Refactoring Software, Architectures, and Projects in Crisis*, John Wiley & Sons, 1998.

[10] Lippert, M., and S. Roock, *Refactoring in Large Software Projects: Performing Complex Restructurings Successfully*, New York: John Wiley & Sons, 2006.

CHAPTER 10

The Offshore Oracle Pattern: Caching Content in Challenged Networks

This chapter discusses a pattern for caching content in a challenged networking environment, with an emphasis on how a DTN can use configurations and metrics for messaging. In unchallenged environments, such as the terrestrial internet, metrics are collected at the endpoints of a communications path. DTNs cannot rely on this approach because of the nature of delayed and disrupted data exchange. Rather than relying on synchronized information across multiple message paths, nodes may choose to send collections of relevant information to a third-party node that can serve as a point of accumulation. The accumulating nodes are termed oracles and they collect and analyze data to provide guidance to other nodes in the network.

These oracles are considered offshore because they exist outside of the scope of a specific messaging exchange. While the metrics that inform oracles comes from messaging exchanges, the oracles themselves can inform network operations, regional asynchronous network management, and local autonomy engines across a variety of networking service types. This pattern can be particularly useful in instances where a DTN experiences periodic or otherwise planned changes to its configurations, such as changes to topology, security, or operational policy.

10.1 Pattern Context

Communicating messages across any network requires coordination of multiple hardware and software resources operating at various levels of service. Coordination, in this context, refers to synchronizing some state between two endpoints of a message exchange. These endpoints can be defined at three different scopes: physical link endpoints, virtual path endpoints, and user application endpoints.

Physical link endpoints exist at either end of a physical transmission medium and represent a single hop in a network. As nodes calculate the overall path of the message through the network, they must reason about both the physical and virtual path of the message. The physical path of a message is the if the ordered set of physical links that the message must traverse between the message source and destination. The virtual path of a message is the set of logical endpoints that a message is sent to at higher layers of a protocol stack. For example, in Figure 10.1, a message may have a physical path of six hops, but the virtual BP path only has

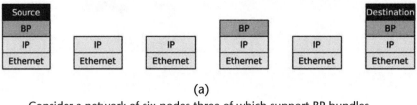
(a)
Consider a network of six nodes three of which support BP bundles.

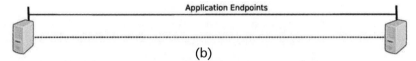
(b)
Application data sees a single hop between the message source and destination.

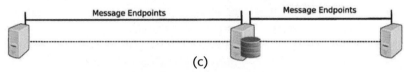
(c)
Bundles see two hops from the source to the next BP agent, and from that agent to the destination.

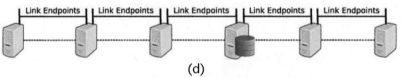
(d)
IP packets and Ethernet frames see all of six hops as they perform the work of delivery across every node.

Figure 10.1 Protocol layers leads to multiple types of endpoints in a network.

two hops. A single virtual path may be satisfiable by multiple physical paths, and individual messages may each take a different physical path along the same virtual path. Similar to virtual paths, application endpoints exist between an application sender and an application receiver. However, in architectures such as DTN that use local caching, store-and-forward, or similar techniques to reduce end-to-end load on a network, the application endpoints may be different from the messaging endpoints.

Except for fire-and-forget networking operations, endpoints perform three basic functions to coordinate information necessary for message delivery. The first function is to establish the ability to exchange data by knowing when physical links are present and when they can be used for transmission and reception. The second is to apply corrective actions to gracefully recover from transient communications errors. The third is to apply preventative actions to reduce the likelihood of future communication errors (see Table 10.1).

Link endpoints coordinate, at a minimum, information necessary to determine access to the shared medium. For wireless links, ALOHA [1, 2] and similar algorithms [3–5] coordinate transmit schedules that minimize interference. For wired links, such as Ethernet, carrier sensing and transmission back-off algorithms

Table 10.1 Examples of Coordinating Information

	Establish Data Exchange	*Corrective Actions*	*Preventative Actions*
Link endpoints	ALOHA channel access Carrier sensing	Automatic repeat requests (ARQs)	Error correcting codes Backpressure mechanisms
Virtual path endpoints	TCP session establishment	ARQs Message reassembly	Congestion measurement
Application endpoints	Application-specific	In-order delivery	Application-specific

deconflict transmitters on the wire [6]. Once media access has been negotiated, data may be lost or corrupted based on transient issues in the medium itself. Corrective actions almost exclusively involve requesting that subsets of user data be retransmitted. For example, ARQs [7] request retransmissions if data does not appear at a receiver within an expected amount of time. Multiple preventative actions are often used to reduce the likelihood of transmission errors. Entropy encoding introduced redundancy in transmitted data [8, 9] so that if there is data corruption at the receiver, the original data has a higher likelihood of being recovered without requiring a retransmission. In addition to oversampling techniques, information can be exchanged relating to data rate and the capacity of receive buffers in the form of backpressure. To exert backpressure on a link, a receiver requests that a transmitter throttles its transmissions to prevent overflowing buffers on the receiver.

Virtual paths coordinate information necessary to measure the effectiveness of the set of physical paths communicating between the messaging endpoints. This includes information such as routing information, security configurations, and congestion measurements. A popular method of coordinating, used in the terrestrial internet, is the creation of a TLS [10] TCP session; once a session is established through a series of round-trip communications, the sender and receiver may begin data transmissions. When messages or fragments are lost or corrupted over the virtual path, retransmissions such as those provided by ARQ schemes may be used. When the receiver over a virtual path represents the destination of user information, corrective actions require waiting for a complete set of user data fragments for both reassembly into a complete message, and if so configured, in-order delivery of those messages to a receiving user application. Preventative measures across virtual paths are difficult to negotiate and involve measuring congestion across preestablished physical paths in the network and restricting access to those paths to avoid congestion altogether.

Application endpoints may require coordination prior to message exchange or may delegate this coordination to the path and link mechanisms provided by the network. For example, applications may require that a TCP session be established, or they may attempt to implement some coordination and stateful information exchange on top of a less reliable protocol such as UDP. If applications communicate over a less reliable medium, then those applications must perform their own error detection and corrective action. In cases where portions of a payload might be useful even if the entire payload was not received (or could not be validated) an unreliable network layer can be used to save resources and complexity. When operating in this mode, applications typically apply their own integrity mechanisms to portions of their payloads so as to identify which portions of their payloads were received uncorrupted.

This practice of fine-grained evaluation of data corruption is commonly used in DTN protocols where the reliability of underlying transport protocols cannot be guaranteed. The BP allows for individual blocks to be integrity-verified, and bundles may be processed even if some blocks are corrupted. Similarly, the Lickliter Transmission Protocol (LTP) [11, 12] allows for the specification of reliable and unreliable data in the same protocol data unit.

Just as corrective actions might be application-specific, preventative actions can be implemented at the application or transport layers. Most often, preventative action involves either sending multiple copies of application data (in anticipation of corruption/loss) or the use of application-specific backpressure mechanisms to avoid congesting links under the assumption that such congestion will lead to corruption/loss. For example, video streaming protocols include backchannels for determining when a transmitting app should buffer data because a receiving application is experiencing resource exhaustion.

As the discipline of computer networking has matured, the number of applications, protocols, configurations, and operational concepts have increased with the variety of networking architectures and characteristics. The amount of coordinated information required to support effective data exchange has also grown and these sets of information span every layer of the networking stack from physical link information to abstract application data representations.

Carrying the volume of information necessary to coordinate information among this variety of endpoints and service types requires the implementation of nontrivial protocols and applications. The design and implementation of these protocols and applications are grounded in the capabilities of the networks in which they are deployed. Specifically, there are a set of assumptions made by these algorithms that are valid in most terrestrial internet applications but invalid in many challenged networking scenarios. Understanding these assumptions, and where they are invalid, is important to determining whether existing information exchange mechanisms are useful or not.

The single most significant assumption made when coordinating between endpoints is that information can be communicated round-trip between these endpoints in an operationally relevant timeframe. Coordination is only effective when the round-trip communication exchanges associated with the link or path can be supported within the performance requirements of the system. On the terrestrial internet, exchanges that occur in tens of milliseconds are unlikely to present synchronization problems. When dealing with interplanetary communications, one-way signal propagation delays of 20 minutes require rethinking negotiation strategies that necessitate multiple round-trip data exchanges prior to passing user data.

The offshore oracle pattern (OOP) provides a configuration and monitoring scheme appropriate for challenged networks that can be used to provide establishment, preventative action, and corrective action information in the presence of delayed and disrupted message exchanges that violate this assumption.

10.2 The Problem Being Solved

DTNs violate the networking assumptions made by traditional endpoint coordination schemes. In such networks, link endpoints can introduce so much signal propagation delay (such as when communicating at interplanetary distances or when communicating over a slower medium[1]) that information cannot be coordinated before it changes. Alternatively, a link can be so disrupted (either because the transmission medium is contested or unidirectional) that requiring any kind of round-trip exchange is impossible (Figure 10.2). Virtual path endpoints can experience all of these problems with no additional insight into network topology, security credentials, or congestion information. In highly delayed environments messages may take days or weeks to be delivered and, in this time, there may be no effective in-band mechanism for assessing the progress of the message through the network. When link and path endpoints fail to provide the types of measurements and coordination necessary to manage data exchange in a network, application endpoints must provide some capability specific to the nature of the network and the data being exchanged.

Rather than relying on in-band methods that coordinate information among and between endpoints, out-of-band mechanisms can provide alternative repositories of preplaced coordination information that can be accessible to endpoints, assist in understanding the effectiveness of unidirectional messaging, and do not require timely, round-trip information exchange.

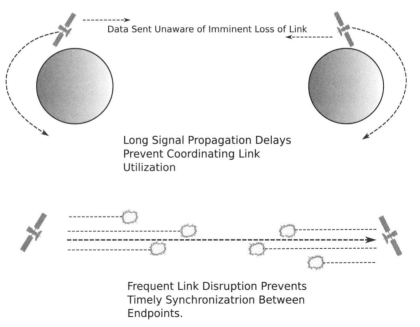

Figure 10.2 Delays and disruptions make endpoint coordination impossible in some DTNs.

---

1. Such as using an acoustic modem, as the speed of sound is approximately 880,000 times slower than the speed of light.

DTNs can benefit from the incorporation of time as an input into the scheduling and configuration of the network. For example, when accounting for signal propagation delays or planned future connectivity, the state of certain networks evolve in a predictable manner over time. Attempting to infer and/or communicate timing information in-band with user messaging may lead to multiple, independent notions of timing among messaging endpoints. In cases where endpoints are separated by the very delays and disruptions being measured it may be impossible to derive this information in operationally relevant timeframes. Some mechanism is needed to generate and communicate coordinated and time-sensitive information that is not dependent upon the links, paths, and applications relying on that information in the first place.

10.3 Pattern Overview

The OOP augments traditional link, path, and application endpoint coordination by accumulating information at special nodes independent of, but accessible to, the nodes participating in user exchanges. These special nodes are considered oracles because they hold information not otherwise available to other nodes in the network. They are considered offshore because they are independent[2] of the link, path, or application endpoints for any user message exchange. This pattern is comprised of four system roles (see Figure 10.3): reporting sources, configuration sources, the oracles, and querying agents.

These system roles are logical elements of an architectural networking design and may or may not translate into individual structural components. Guidance on mapping pattern roles to deployed software applications is provided later in this chapter.

10.3.1 Role Definitions

Reporting sources (RSs) generate the data necessary to evaluate and adjust messaging in the network. These sources are those endpoints (link, path, and application) and waypoints that directly measure performance and generate metrics for user messaging. RSs send their data to one or more oracles to be analyzed and fused with other information and ultimately redistributed. The ability to fulfill the RS role may be built into supported protocols or may require dedicated application software to collect and transmit this information. For example, a node may have an RS for each radio and an RS for each networking protocol.

There are a variety of ways in which the role of RS can be translated into one or more software applications. One approach is to generate a one-to-one mapping of RS application to local information source and have each RS communicate directly to oracles. Another approach is to define a single RS application for each node and have all local information funneled through that RS to provide a single point-of-contact for information sent to oracles. A third approach is to balance the number

---

2. Independence in this context means not dependent on user messaging. This is not meant as a requirement that the oracle nodes and messaging nodes always exist on separate hardware or exist in separate parts of the network.

10.3 Pattern Overview

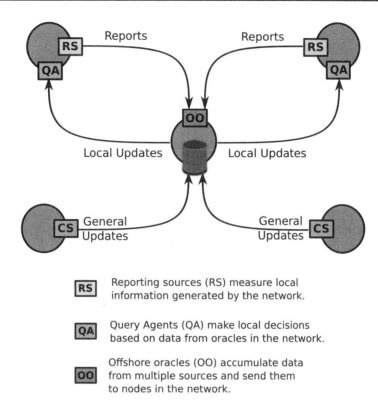

Figure 10.3 Roles in the OOP.

of RS applications on a node using some criteria such as one RS application per service type (Figure 10.4).

Configuration sources (CSs) provide information generated independent from the messaging endpoints being managed by this pattern. Unlike RS applications, CS applications typically provide information with a scope broader than a single messaging path and include information such as external observations, scheduled networking events, requests for new configurations, or other state changes asserted by network operators. Another difference between RS and CS roles is that CS roles may include information about future events whereas RS roles only measure and report on that which has already happened. The architectural approach to mapping CS roles with software applications is similar to the approaches available for RS: one-to-one by configuration item, single application for a single configuration source node, or some balance based on user or network specifics.

Offshore oracles (OOs) collect information from RSs and CSs, analyze that data, and provide it back to nodes as needed. OOs exist as persistent roles in a network and are not instantiated for any specific message exchange. The information they collect from multiple sources is used to generate predictions about the future state of the network. In cases where configurations come directly from network operators, these predictions may have a very high confidence associated with them. In cases where elements such as link state or congestions are inferred from past measured performance, the predictions may have less confidence associated

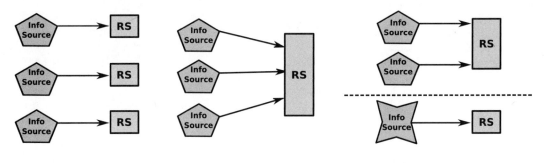

| One RS application can be built for each information source on a node. | A single RS application can collect information from all other information sources. | RS applications can be allocated to information sources based on common traits - such as service types. |

Figure 10.4 RS applications can be deployed in a variety of ways on a node.

with them. OOs communicate their information to those Querying Agents that it believes would benefit from this predictive information.

The mapping of OO roles to application software is slightly more complex than the mapping of RS and CS roles. As a type of content cache in a network, specific nodes in the network should be engineered to have the resources necessary to accumulate and analyze information and to exist in appropriate spots in the network that can influence message exchanges without being solely reliant on those message exchanges. Since every node in a network participates in some message exchange, it is likely that OO and RS applications will sometime be colocated. In some cases—as a function of resources and network placement—an OO application can also be colocated with a CS application.

Query Applications (QAs) request and cache information from oracles to help their node with current and future message exchange. QAs request information relevant to the endpoints resident on their node. For example, QAs may receive information that helps video streaming applications determine whether and when to stream video information over unidirectional links. Alternatively, they may help build contact schedules for routing scheme agents or set configurations for software-defined radios. QAs have special roles in this pattern, as they directly influence when generated data should be placed into the network and whether missing data needs to be retransmitted. The mapping of these roles to software applications follow the same guidance for RS and CS roles.

10.3.2 Control and Data Flows

Data flows among components in this pattern are one-way, consistent with the observation that this pattern is used as a substitute for round-trip data exchange (see Figure 10.5). Neither the control nor the data flows assume that communication to or from an OO occurs within an operationally relevant timeframe. In cases where multiple roles are coresident on the same device, communications will be timely and reliable. In cases where roles are distributed across long propagation delays or in partitioned subnetworks, communications may be lagged and/or unreliable.

10.3 Pattern Overview

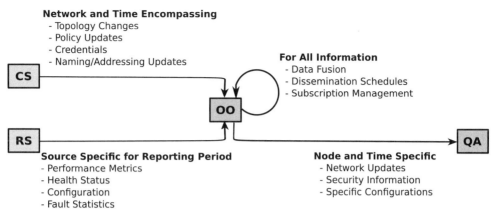

Figure 10.5 Data flows through the OOP.

In practical deployments, the packaging of pattern roles into software applications will be driven by the resource characteristics of the node and the administrative roles and permissions deployed in the network. Just as RS, CS, and QA roles may be implemented in multiple or single applications, combination applications fulfilling multiple roles may also be developed. In all such combinations, it is likely that a local node will keep a local cache of information (either in-bound from oracles or outbound to oracles) to help buffer transient issues with the network itself.

The number and positioning of OOs in the network can have an impact on the nature and reliability of data flows in the network. Because oracles receive both measured data (from the past) and inferred data (for the future) they have a sense of the predicted change of the network over time. In cases where multiple OOs exist in a network, keeping them coordinated may not be possible—in the same way that coordinating other endpoints in a challenged network may not be possible. Therefore, OOs should be deployed so as to have the most beneficial impact in areas of the network where they are likely to be able to sustain timely information. The deployment of OOs to transition points in a network increase the likelihood that the information kept at the OO is relevant to its immediate domain without needing complete synchronization for others OOs in the network.

Certain flows in this pattern are data driven, and others are time-driven and the distinctions are driven both by the nature of the data and the architecture of the network. For example, measurement information collected by RS applications often happens periodically over time, and the transmission of that information—individually or as a cached set—can happen as a function of elapsed time, quantity of data, or availability of contacts. CS information, representing new configuration information for the network, may be data-driven when new configurations are available or time-driven based on planned changes to the network. QAs may either request information from oracles as they need it (using a pull mechanism) or subscribe to oracle information as it becomes available (using push mechanisms). As information is produced, these agents update the configurations of the local node to help with future messaging exchange.

10.4 Service Types

This network architecture pattern can be used with any service type as OOs store information across the networking service spectrum (see Table 10.2).

Transport services act as RSs within this pattern. Software and firmware operating at this layer provide the lowest level of insight into the individual links in a network, including metrics such as ranging, throughput, signal strength, link quality over time, and where applicable, scheduling of link establishment. In cases where an individual node supports multiple transports (such as Ethernet, Wi-Fi, Bluetooth, near-field communication, cellular) the information provided to OOs also helps higher layer services make determinations relating to routing, congestion, and throughput. Transport services never act as their own OOs or QAs as transport services focus only on the immediate incoming and outgoing links for which the node has authoritative information. They do not act as their own CSs as any out-of-band configuration occurs as part of physical network configuration and any in-band configuration/negotiation occurs outside of the context of this pattern.

Networking, federating, and overlay services report on their respective networking functions. Type II services report on the configuration and performance of the physical network, to include security configurations, routing information, naming and addressing, topological changes, and network management metrics. Type III services report on the number and types of federated networks and the configurations necessary to translate information among them. Type IV services report on the same information as Type II services, but for the overlay network and not the physical network.

Similarly, networking, federating, and overlay services each provide their own OOs holding information for their specific service types. This is necessary because different service layers do not have access to the same set of nodes, naming, or addressing. For example, an OO instantiated in a networking overlay may not be reachable by applications running at the physical networking layer. Conversely, and by design, services that are instantiated at the physical networking layer are hidden by services operating in a different federated network or in an overlay. Since OOs may be instantiated at multiple service layers, their QAs and CSs must also be instantiated at multiple service layers.

Finally, user applications neither act as RSs nor OOs. In this pattern, reports flow from an RS application to an OO application. If a user application were to have one endpoint as an RS and another endpoint as an OO, it would be equivalent to messaging user information between user application endpoints—a feature assumed impractical when applying this pattern. Instead, user applications may fulfill

Table 10.2 Offshore Oracle Service Types

Service	RS	OO	QA	CS
I. Transport	X			
II. Networking	X	X	X	X
III. Federating	X	X	X	X
IV. Overlay	X	X	X	X
V. User endpoint			X	X

the roll of QA and query OOs about information to make data production and consumption decisions (Figure 10.6). Additionally, user applications may serve as CSs by which underlying networks can be configured as a function of the kind of data that may be generated by a user application. In this way, user applications may analyze and alter information in available OOs, but with a focus on understanding the underlying network and not as a mechanism for passing additional user data.

10.5 When and How to Integrate

The successful use of this pattern is predicated upon multiple assumptions on the architecture of the network and the types of data that flow through the network. This section describes the times when this pattern should and should not be used and how to make design decisions on the mapping of pattern roles to software applications.

10.5.1 When to Use This Pattern

This pattern is useful when end-to-end link status cannot be measured and configured in real-time. Instead of performing in-band configurations, third-party nodes

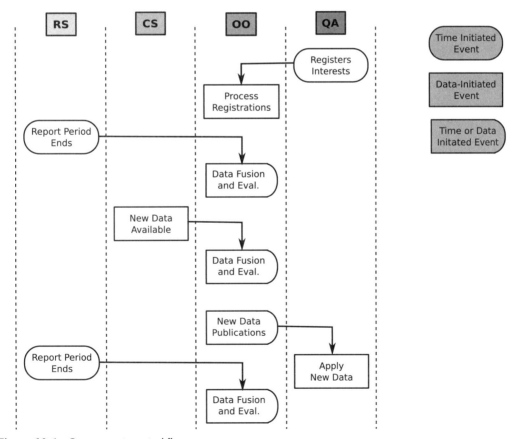

Figure 10.6 Component control flows.

collect individual sets of information that, in aggregate, yield insights into the future operation of the network. The OO pattern should be considered whenever any of the following conditions are true about the network.

Endpoints cannot coordinate information in real time. The purpose of OOs is to hold information about message exchange in cases where the traditional, in-band mechanisms will not work. If endpoints (at any level) cannot exchange information in a rapid and reliable manner, then traditional algorithms used to adjust messaging parameters cannot be used. Coordination becomes an open-loop control problem enabled by caches of networking data rather than a closed-loop control problem enabled by timely round-trip data exchange.

Configurations cannot be synchronized across the network. The placement of OOs only help when information cannot otherwise be propagated across the network in operationally relevant timelines. When nodes (or partitions) are disconnected for extended periods of time, they have no timely access to configuration information. In these cases, OOs can help by caching future configurations and other predictions for use during periods of node or partition disconnection.

Past performance can predict future performance. Often the things that challenge network performance are based on periodic phenomena. In space-based systems, planetary motion and spacecraft orbits challenge networks in predictable ways. In terrestrial networks, environmental issues, power-cycling, and geocaching can similarly alter information coordination in predictable ways. Because OOs collect information from multiple sources and fuse their data to provide updates and guidance to QAs, they are excellent resources for pattern analysis and predictive analytics in a network.

Appropriate waypoints exist in the network. If every node in the network needs to become an OO to carry information, then the pattern becomes impossible to implement. Establishing OOs requires that there exist nodes in a network that can both collect and distribute information in a useful way. This means that OO nodes have the compute and storage resources necessary to process and fuse information and that these nodes have sufficient connectivity with other nodes in the network (or local network partition) to be able to share the information in relevant timeframes.

10.5.2 Recommended Design Decisions

The detailed design of components, and the way in which they should be integrating into a functioning network, are specific to the eventual operational environment. However, in generating the detailed designs of these components, multiple recommendations exist for how to make design decisions that result in a scalable, efficient, and operational implementation.

Minimize the number of applications on a node. There are many distinct sources of information related to messaging across all service types on a networking node. Building one application per role implies dozens of applications running on a node using significant memory and computational resources. The OO pattern is deployed in challenged networks whose nodes are often resource constrained and unlikely to support the inefficiencies inherent in supporting dozens of applications providing similar functions. Wherever possible, independent sources of information should collect and report their information through the smallest number of applications.

Maximize use of shared storage. The insight driving the OO pattern is that independent pieces of information, collected over time, provide information about the current and future state of a network. This analysis requires access to all of the information so that it can be appropriately annotated, searched, and fused. Shared storage structures such as databases, caches, and shared memory segments provide normalized access to data in support of these predictive analytics.

Prefer push notifications. Pull requests require round-trip communications which are assumed to be unreliable and not timely in networks using this pattern. RS and CS applications should push information to known OOs, and OOs should push information to QAs so as to not miss out on any connection opportunities in the network. When using store-and-forward protocols such as the BP, push opportunities can simply result in sending information into the network when it is available an letting the underlying protocols handle delivery.

Minimize the number of messages. Just as minimizing the number of applications reduces memory and computational overhead, minimizing the number of messages in the network reduces the network overhead associated with this pattern. Reducing messaging overhead reduces the amount of space needed to store-and-forward messages, reduces the amount of time necessary to transmit the information, and reduces the likelihood that messages with useful dependencies would not be uniformly received when transmitted individually. Reducing the number of messages should be a strategy even when it means combining information from multiple sources and across multiple service types.

Annotate data with relevant time information. OOs act as centralized repositories of information collected at various times through various parts of the network. The information coming in to an OO might not have been generated close in time to other received information. Data from RS applications may have been generated recently or (depending upon delays and disruptions) long ago. Information from CS applications may reference far-future configurations. All information into, and out of, an oracle should be annotated with multiple timestamps, to include the time at which the information was generated and the timespan for which the information should be considered relevant.

Place oracles at gateways and on data mules. Gateway nodes represent the interface between two or more portions of a network that operate differently. These differences can be syntactic (such as the use of different physical links or different protocols) or semantic (such as at the intersection of a challenged/unchallenged or trusted/untrusted subnetworks). To the extent that OOs collect information associated with messaging, gateways provide an opportunity to locate information in a place that is likely central to most information paths. Data mules represent areas where multiple messages pass through a single node making them natural places to collect and distribute information.

10.6 What Can Go Wrong

There are very few guarantees in highly challenged networks, making the deployment of this pattern susceptible to a variety of potential issues.

Insufficient connectivity among roles. The data and control flows for this pattern operate in conditions where reliable messaging does not exist. Protocols such as the BP might need to store information for very long periods of time prior to communication. If OOs cannot receive, analyze, and communicate information then they cannot provide useful information for future messaging attempts. There is no protocol or pattern that can communicate information absent a data link.

Resource exhaustion on the oracle. The amount of information available to OOs will be driven by the size and diversity of the network, the data volumes communicated through the network, the timeframes associated with information, and the number of OO applications deployed. This may place large resource demands on OOs to process and store vast amounts of information—particularly for networks where nodes may be disconnected for very long periods of time.

Incorrect inferences. OOs provide predictions and status based on both measured performance and asserted configurations. These predictions may be reliable for situations where patterns exist and are discoverable in the OO's data. However, they may be unreliable in situations where there are no patterns in the data or an insufficient amount of data has arrived at the OO to support this analysis.

Out-of-sync oracles. When multiple OOs are deployed in a challenged network, they will receive information and configurations at different times, leading to different information provided to different QAs. When OOs in the network are not (or cannot be) synchronized, individual QAs may receive different and potentially conflicting configurations.

10.7 Pros and Cons

The use of the OO pattern can be a powerful tool to coordinate configuration information and help anticipate issues with messaging, although this requires authoring multiple new applications and ensuring that networking nodes maintain appropriate resources. This section summarizes some of the benefits and pitfalls of using this pattern in a network (see Table 10.3).

10.8 Case Studies

This section provides three examples of using this pattern applied to three different service types. Status reporting for the BP is an example of a minimal set of information used for overlay services. Network topology management provides an example of applying the pattern to a network service. Security policy updates demonstrates ways to apply the pattern for a user application.

10.8.1 BP Status Reporting

The BP mechanism for sending administrative records fulfills the responsibilities of an RS application and the overall use of this mechanism for network support conforms to the OO pattern. Any BPA, in accordance with policy, can generate a special type of administrative record called a status report. Status reports signal the occurrence of specific bundle processing events on the BPA and they are sent to

Table 10.3 OO Pattern Pros and Cons

Context	Pros	Cons
Applying configurations	Oracles provide a central point of entry for configuration sources.	Multiple oracles may be out of sync.
Performance measurement	Oracles provide centralized stores for collecting and analyzing performance measurements.	Performance measurements may be received asynchronously making it difficult to build a common operating picture.
Predictive analytics	Patterns in the data can be discovered when data is co-located.	Asynchronous updates of oracles can prevent patterns from being discovered.
		Processing may not help messaging exchange when there are no patterns in the data.
Cost and complexity	Oracles provide single points of update and sampling for network management.	Oracles require additional resources at the nodes they occupy.
	Applications can be developed without reinventing mechanisms to coordinate information.	Additional network architecture complexity to determine where to place oracles and how to maintain applications.
Messaging support	Oracles provide information that cannot be negotiated in-band on an endpoint-by-endpoint basis.	Oracles may not be immediately accessible to an endpoint.

special report-to endpoints. These report-to endpoints may be unrelated to the message path of the bundle for which the report was generated. Each BP status report contains information about the bundle (such as its source node ID and its creation timestamp) as well as information about the initiating event.

BPv7 defines four events for which a BPA may generate a status report: bundle receipt, bundle forwarding, delivery of a bundle's user data to a local application, and deletion of the bundle. Additionally, each status report can include a reason code to provide additional context for what may have triggered the event (see Table 10.4).

Table 10.4 BPv7 Status Reason Codes

Code	Meaning
0	No additional information
1	Lifetime expired
2	Forwarded over unidirectional link
3	Transmission canceled
4	Depleted storage.
5	Destination endpoint ID unintelligible
6	No known route to destination from here
7	No timely contact with next node on route
8	Block unintelligible
9	Hop limit exceeded

The mapping of status reporting mechanisms to OO pattern objects is relatively straightforward (see Figure 10.7). BPAs serve as RSs that generate administrative records and send them to report-to endpoints for the bundle. The report-to endpoints function as the OOs as they accumulate information from multiple BPAs and multiple message paths without the requirement that that exist within the message path of the bundle. This use case does not specify CS or QA roles other than to imply that the collection of status reports is used to update configurations through some network management function.

10.8.2 Topology Management

Topology management informs nodes of the structure of the network, including predicted future contact opportunities and performance statistics for past contacts. This is a particularly difficult task in a dynamic and challenged network where it may be difficult or impossible to globally update topological information prior to it changing. In this pattern, all nodes that have authoritative information on their own contacts (measured or predicted) can act as RSs in cases where they measure topology and as CSs in cases where they predict topology.

Consider an example of an Earth-to-Mars internetwork consisting of a mission operations center, deep-space network, Mars-orbiting spacecraft, and a series of Mars landers. In this example, every node can act as an RS for topology as every node has one or more contacts passing traffic (see Figure 10.8). These statistics can be sent to OOs to provide confidence levels associated with link quality and accuracy of contact start and stop times. Every node can also serve as a QA because every node has a need to understand relevant portions of the network topology.

Nodes that predict future contact opportunities serve as CSs in the system. A mission operations center, which commands various telecommunications systems,

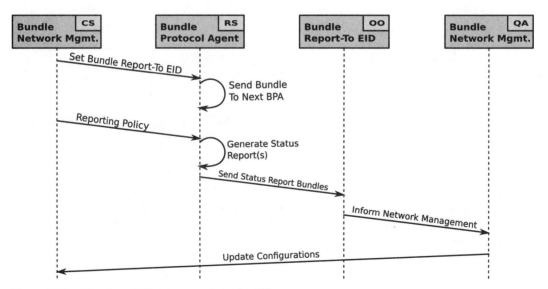

Figure 10.7 Mapping of BP status reports to the OO pattern.

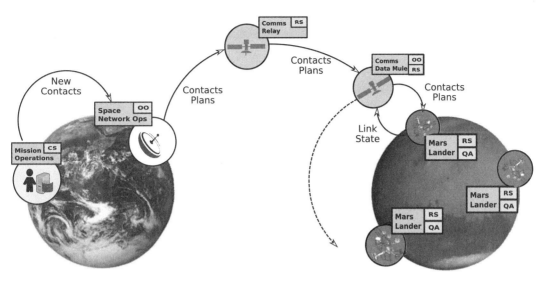

Figure 10.8 Topology management in an Earth-to-Mars network.

may understand planned network outages or other topology changes. An orbiting spacecraft may autogenerate predicted contacts over an upcoming time period as a function of its orbit and the position of other nodes. OOs exist at gateways and on data mules. OOs at deep-space network nodes bridge the terrestrial and challenged portions of the network. OOs on Mars orbiters act as data mules for various Mars landers and provide local information absent direct-to-Earth contacts.

10.8.3 Security Policy Updates

Security policy updates are difficult to negotiate in a challenged network—the lack of timely round-trip information exchange precludes the use of several proven security techniques such as streaming-mode ciphers and synchronized certificate authorities. An example of a challenged network in which security policy updates benefit from this pattern is a sensor network deployed to a remote location with irregular connectivity back to the terrestrial internet via data mules. By using this pattern, OOs can be configured with a series of security policies to enact over time, and those policies can be enacted on various nodes in the sensor network even when the sensor network is disconnected from the terrestrial internet.

In this example, user applications generate policy updates, such as generating new long-term keys, determining what cipher suites are used in what conditions, and mandating behavior in error conditions. In this way, user applications act as CSs generating configurations and their usage over time. Any security processing node serves as an RS providing security-related metrics, such as which long-term keys are in use, how often they have been used, and for what period they have been active. OOs exist on data mules to communicate configurations to partitions disconnected from CSs. Nodes in these partitions contain QAs and receive security updates, as determined by the OOs.

10.9 Summary

Coordinating information across messaging endpoints in a challenged network is difficult and rarely accomplished in real time. Methods used on the terrestrial internet to measure performance and adjust performance in real-time fail when there is no guarantee of timely round-trip data exchange. An alternative to closed-loop control of messaging endpoints is open-loop control information by special nodes in the network that opportunistically accumulate information. The benefit of the OO pattern is that is does not require timely round-trip data communication to function; OOs collect, fuse, and distribute information as best as possible and when contacts allow. OOs work best when they are positioned strategically at gateways or other boundaries within a network. This positioning allows them to be near most messaging in the network to collect the most information possible. Additionally, this positioning allows OOs to provide information to isolated partitions in the network for which they act as a gateway.

This pattern is best used when the information stored on OOs is predictive in nature, allowing for extended periods of disconnection from CS applications but still able to communicate relevant information to QAs. The pattern is only as good as the connectivity which supports it, as information cannot be exchanged without a physical link and unless there are underlying patterns in the data the predictive analytics on an OO will not yield desirable results. However, for the instances where the pattern applies, this can serve as a useful framework for open-loop control of a network at varying levels of service from individual transport-layer concerns to higher-level user applications.

10.10 Problems

10.1   Discuss why each type of endpoint (link, virtual path, and user application) requires in-band, round-trip communications. What happens to messaging across these endpoints as signal propagation delays grow?

10.2   A virtual path is defined as the set of all physical paths used to communicate messages from the virtual path source to the virtual path destination. Explain why metrics across such a virtual path inform message exchange even though the virtual path is comprised of multiple physical paths.

10.3   Provide three examples of configuration information that can be preplaced on an offshore oracle and communicated to query agents at specific time intervals.

10.4   Should an OO push to a QA all information it has relative to that QA, or only information within a specific timeframe? Defend your answer.

10.5   Describe a design rationale for mapping RS roles on a node to more than one RS software application running on the node. Explain under what circumstances this design be necessary.

10.6   Provide an example of a CS that does not involve a human operator determining and injecting new information into a network.

10.7   Describe why QAs should receive push notifications from an OO rather than having a QA pull information from its OO.

10.8   Assume you are designing a QA application for a node which may receive push notifications from multiple OOs. Describe three reasons why two different OOs may send conflicting information to the QA. Describe an approach to deconflicting such data.

10.9   Describe a strategy to avoid resource exhaustion on an OO node. Explain under what circumstances this would result in incomplete information being sent to QAs.

10.10 Describe how new OOs might be added to an existing network deploying this pattern.

10.11 Describe a method by which multiple OOs can synchronize their data.

References

[1]   Abramson, N., "THE ALOHA SYSTEM: Another Alternative for Computer Communications," *Proceedings of the November 17-19, 1970*, Fall Joint Computer Conference, ACM, 1970.

[2]   Crowther, W., et. al., "A System for Broadcast Communication: Reservation-ALOHA," *Proc. 6th Hawaii Int. Conf. Syst. Sci.*, 1973.

[3]   Baccelli, F., B. Blaszczyszyn, and P. Muhlethaler, "An Aloha Protocol for Multihop Mobile Wireless Networks," *IEEE Transactions on Information Theory*, Vol. 52, No. 2, 2006, pp. 421–436.

[4]   Lee, S.-R., S.-D. Joo, and C.-W. Lee, "An Enhanced Dynamic Framed Slotted ALOHA Algorithm for RFID Tag Identification," *The Second Annual International Conference on Mobile and Ubiquitous Systems: Networking and Services*, IEEE, 2005.

[5]   Zhen, B., M. Kobayashi, and M. Shimizu, "Framed ALOHA for Multiple RFID Objects Identification," *IEICE Transactions on Communications*, Vol. 88, No. 3, 2005, pp. 991–999.

[6]   Harrington, J. L., *Ethernet Networking*, Academic Press, 1999.

[7]   Lin, S., D. J. Costello, and M. J. Miller, "Automatic-Repeat-Request Error-Control Schemes," *IEEE Communications Magazine*, Vol. 22, No. 12, 1984, pp. 5–17.

[8]   Médard, M., and A. Sprintson (eds.), *Network Coding: Fundamentals and Applications*, Academic Press, 2011.

[9]   Heegard, C., and S. B. Wicker, *Turbo Coding*, Vol. 476, Springer Science & Business Media, 2013.

[10] Rescorla, E., *SSL and TLS: Designing and Building Secure Systems*, Vol. 1, Reading: Addison-Wesley, 2001.

[11] Burleigh, S., M. Ramadas, and S. Farrell, "Licklider Transmission Protocol-Motivation," *IETF Request for Comments RFC 5325*, 2008.

[12] Ramadas, M., S. Burleigh, and S. Farrell, "Licklider Transmission Protocol-Specification," *IETF Request for comments RFC 5326*, 2008.

CHAPTER 11
The Training Wheels Pattern: Open-Loop Control

This chapter discusses a pattern for allowing applications to operate independently when disconnected from remote sources of information in a network. Applications operating in unchallenged environments can use closed-loop control mechanisms that include applications running on remote nodes to make processing decisions in real-time as part of the normal function. Achieving this closed-loop control requires timely, uninterrupted access to necessary data caches and remote services. This type of timely, reliable remote access cannot be guaranteed when operating in a challenged environment, such as a DTN, so closed-loop control cannot be used remotely in these situations. An alternative to closed-loop control is the open-loop control concept where applications make decisions using only the (potentially incomplete) set of information known to the local node at the time a decision needs to be made.

The training wheels pattern (TWP) provides a set of utilities that help networked applications make decisions using only state information known to the local node. To accomplish this, the TWP is designed to be a monitor-response autonomy system, but one focused on helping applications maintain existing function; not determining new courses of action such as would be the case with artificial intelligence or other machine learning approaches. Just as training wheels reduce the likelihood a bicycle will fall over when ridden by a novice, this pattern provides a mechanism to preserve existing application utility in the absence of expert control.

11.1 Pattern Context

Developers make assumptions to simplify the design and implementation of their applications. These assumptions span multiple domains, including those about user inputs (types of requests, data volumes), those about the network (security, timeliness), and those about the operational environment (frequency and type of stimuli). Assumptions can be differentiated based on how and how often they must be validated. Some assumptions can be incorporated into the design and implementation of the application and its environment such that the assumption can never be invalidated once the application is deployed. Other assumptions must be revalidated over time as the application operates in the network.

A static assumption needs to be verified a single time, as its validity will not change as a function of user, network, or environmental changes. For example, an application may be designed with the static assumption that the underlying network provides a secure environment. In this case, the application might not include any independent security services in its implementation. Were that assumption to no longer be the case, a new version of the application would need to be written, but that would involve a new development effort around a new set of assumptions.

Certain types of static assumptions (such as the presence of security services in an underlying networking layer) can be built into the design and implementation of the application itself. Alternatively, applications might check a static assumption once at start-up, such as part of run-time configuration. Because static assumptions are often resolved by the time an application is up and running on a node, they rarely are affected by whether the application is running in a challenged or unchallenged networking environment.

A dynamic assumption must be verified more than once, as their validity changes as a function of user, network, and/or environmental changes. For example, a security-aware application may assume that credentials used as part of its security processing have not expired. This is not a static assumption, as there may come a time when the credentials have expired over the operational lifetime of the application.

Applications must validate dynamic assumptions whenever those assumption may have changed. These assumptions can be validated by the application, by another application on the same node, or by some other application on some other node in the network. In unchallenged networking environments, a common architecture is to locate validation services on well-resourced nodes in the network and to have network applications use those services as necessary to validate their own assumptions when necessary (Figure 11.1). For example, prior to using a security certificate, terrestrial internet applications will first interact with a remote certificate authority to ensure that existing certificates have not expired. Unlike a static assumption, applications must determine when a dynamic assumption must be validated. Performing validation too often incurs computational overhead and

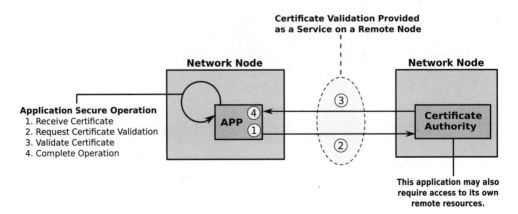

Figure 11.1 Applications use network services for complex operations.

processing delays. Performing validation not often enough risks using incorrect data as part of an operation.

Determining how often a dynamic assumption must be validated will be influenced by the specificity of the assumption itself. The more specific the assumption, the more unambiguous its method and frequency of validation. A rubric for determining the appropriate level of specificity would benefit both programmers implementing validation procedures and network architects designing network services. A rich set of such rubrics exist [1–3] for evaluating software requirements, with one popular method being to make requirements specific, measurable, achievable, relevant, and timely (SMART). The SMART rubric does not need to be constrained to the domain of software requirements; as a method of unambiguously defining information it can also be adapted to evaluate dynamic assumptions. This does not imply that a dynamic assumption is a software requirement. Software requirements verify and validate an application prior to its operational deployment. Dynamic assumptions are checked by deployed applications to test the processing environment prior to performing a function. However, if assumptions can be specifically expressed such that whether their criteria are met can be measured in a timely and relevant way, then software can be engineered to determine that the network is operating within that bounds of that assumption.

Of the SMART criteria, specificity and measurability are the most important as they speak to removing ambiguity. A verifiable requirement, at a minimum, must be clearly and unambiguously defined in such a way that adherence to the requirement can be measured. Dynamic assumption validation can be defined as the process of first measuring the processing environment and then evaluating measurements against the criteria specified in the assumption. By definition, a specific, measurable assumption is one that can be validated.

Consider the dynamic assumption that the application will always have a valid security certificate, which is neither specific nor measurable. It is unclear what the terms always and valid mean, and whether this applies to a single certificate, a special set of certificates, or all certificates used by the application. Enforcing this assumption risks subjective interpretation and differing behavior over time or among interoperating applications. An alternative, specific, measurable formulation of this assumption would be that security certificates would be considered valid by a certificate authority at the time of their use. This formulation is both specific and measurable: the assumption holds for all certificates used by the application; the certificates must be valid at the time of their use; certificate validity is determined by a certificate authority.

In cases such as the security certificate example, applications coordinate with other services—either on the same node or elsewhere in the network—to understand whether dynamic assumptions are being met. This availability of such services is itself a dynamic assumption and one that is not always valid in a DTN. By clearly stating the dynamic assumptions in a system, it may become clearer to both network architects and application engineers whether these assumptions are appropriate and deployable in an operational system.

The TWP provides a mechanism for applications to validate dynamic assumptions made by applications in the network and to gracefully react to conditions where those assumptions cannot be met.

11.2 The Problem Being Solved

Networked applications, by definition, exchange data across a network. Some of those exchanges are related to the purpose for which the applications exist such as sensor measurements, user inputs, application commanding, and commanding responses. Some exchanges are related to helping the application fulfill its purpose and are never seen by the application users. Exchanges relating to user data occur over the data plane and exchanges relating to helping keep the application functioning occur over the control plane. In some systems, the data and control planes are physically separated (such as using different radios), in other cases the planes are logically separated (such as through different sessions over a single physical channel), and in other cases the planes are intermingled and only identifiable by the type of messages received over a single interface (see Figure 11.2).

The types of mitigations that can be applied to data planes are not always useful when applied to control planes. For example, store-and-forward and other time-variant mechanisms are useful for ensuring eventual operation of the data plan in DTNs. They allow user data and user commands to progress through the network as patiently as possible, particularly when data remains relevant for longer time periods.

This same approach does not always work as well for the control plane. When challenges in a network prevent the timely access to control information then the application may not have the information necessary to proceed with an operation. If the control of the application is interrupted, it may not be able to generate user data (as would be communicated over the data plane) and would therefore be unable to fulfill the purpose for which it exists in the network.

Because of the importance of the control plane for the proper function of an application, there needs to be some way of providing a local control plane that can be used when the nominal control plane for an application is unavailable. The local control plane could cache information and provide guidance based on the last known set of information available to a local networking node and be used in lieu

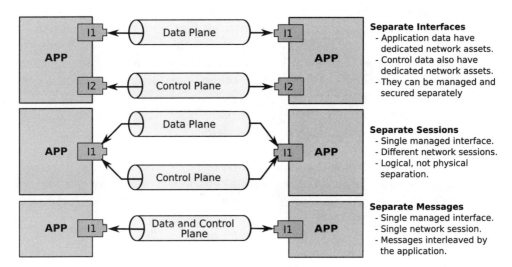

Figure 11.2 Control and data planes can be implemented in a variety of ways for an application.

of other information when the local node is disconnected from more authoritative sources in the network. The benefit of this approach is its reliability; the local control plane would never be offline because it is coresident with the applications it serves (see Figure 11.3). The risk of this approach is that local caches of remote information can be out-of-date if not regularly refreshed.

The type of information that can be provided by a local control plane usually involves information generated on the node itself, and cached information retrieved from messages passing through the node, including information received from various remote control planes in the network. In cases where application dynamic assumptions can be validated using this type of information, a local control plane can be a deterministic way to implement a local autonomy system to help applications function in times of poor connectivity.

11.3 Pattern Overview

The TWP provides a local control plane that applications on a node can use to validate dynamic assumptions and adjust configurations to help maintain the validity of those assumptions. The pattern provides constrained autonomous behavior on a node without requiring applications to operate autonomously all the time, or in ways which make its actions unpredictable or insecure. The training wheels provided are the actions taken by the node to keep assumptions validated (if possible) and to prevent cascading errors caused by applications operating using unexpected data or with an unexpected environment. The pattern is comprised of six logical components: local data collectors (LDCs), remote data aggregators (RDA), a data derivation engine (DDE), a data repository (DR), a predicate evaluator (PE), and a local control plane (LCP) (see Figure 11.4).

These roles represent logical functions that must be fulfilled to achieve the features of the pattern. As logical constructs, they may not have a one-to-one mapping to concrete components of a software design. For example, a role may be implemented by multiple software functions (or software applications) working togeth-

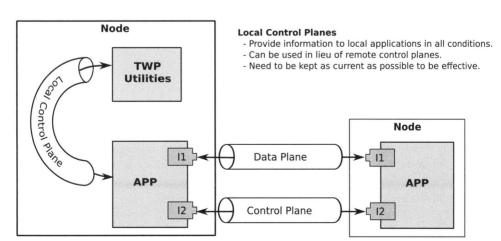

Figure 11.3 Applications in DTNs use both a local and remote control plane.

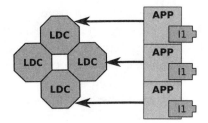

Local Data Collectors
- Sample data from local sources
- Many-to-many mapping to sources

Data Derivation Engine
- Derives new data values from existing data values.
- Like the LDC, the DDE is a source of data for this pattern.

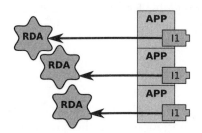

Remote Data Aggregators
- Sample data from remote sources
- Examines data sent to this node
- Examines data sent through this node.

Predicate Evaluator
- Evaluates predicate expressions
- Used to determine if an assumption is valid.

Data Repositories
- Identify data needed for the pattern.
- Indexes data for fast retrieval.
- Annotate data with time and source.

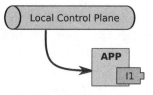

Local Control Plane
- Issues controls in the absence of a remote control plane.
- Helps self-configure node to maintain operational assumptions

Figure 11.4 System roles of the TWP.

er, representing a one-to-many relationship between roles and implementations. Similarly, a single software application might perform multiple roles.

11.3.1 Role Definitions

LDCs exist on the same node as the application(s) that make use of the TWP and sample those data necessary for validating application assumptions. LDCs exist in a many-to-many relationship with information sources on a node. In this way, LDCs are conceptually like RSs in the OOP. Examples of the kinds of data generated by LDCs include information collected from imagers and sensors, health information such as CPU utilization and temperatures, and time information from clocks and other oscillators. There are often multiple such collectors because they can be added to nodes over time as new applications are installed or new instrumentation is required to implement this pattern for different applications.

Remote data aggregators (RDAs) collect information generated remotely by applications not resident on the node. This includes both information directly addressed to the node and other network information that passes through the node. Directly addressed information includes lookup from authoritative sources elsewhere in the network, directly pushed information from network operators, and general control plane information from the set of applications resident on the node. Information from messages passing through the node can include deep-packet inspection of messages. For example, when using BP in a network, a node may evaluate certain types of extension blocks to derive information about upstream nodes or to better predict congestion at downstream nodes. In these cases, BP bundles might not be directly addressed to the current node but may be stored in such a way that allows their data to be inspected and used to update DRs on the node.

DRs store and index pattern information known to the node. They serve as a local cache providing the data needed to validate application assumptions. Importantly, DRs provide information in a way consistent with naming conventions and inclusive of the metadata necessary for lookup by other pattern roles. DRs must perform four major functions: data identification, storage, indexing, and annotation.

Data identification refers to the need to determine which data must be collected to validate assumptions for applications on the node. Attempting to store all data generated on a node will exhaust node computation and storage resources and is generally not scalable over time. TWP DRs only store information necessary for use by the TWP. Storage may refer to the copying of data from data sources into a separate cache or database or the keeping of references into the original data locations. Regardless of the storage mechanism, high-performance access to the data is required to reduce the overall amount of time needed to implement the pattern responsibilities. Data indexing refers to the means by which other parts of the TWP identify and request data values accessed through the DR. Finally, DRs must annotate their data to include information such as the original data source and the time when the data was collected.

The convergence of information into a DR provides a convenient mechanism for standardizing information that can be generated from multiple sources at multiple times (Figure 11.5). In this way, the DR almost acts as an OO, except that it is fed from, analyzed, and used entirely by the local node. In doing so, DRs must make a variety of decisions typically associated with data caches to include how to

- Determine when a data value can be removed from the repository;
- Derive unique values from existing values;
- Resolve conflicting values from the same datum from multiple sources;
- Identify and correct for data values that are missing or have gone stale.

A DDE generates new data derived from the values of existing data in the DR. The DDE is a special type of LDC that treats the DR as its data source. Examples of derived values include averages of samples over time, counts of items received, and new values calculated as a function of other collected values.

A PE engine accepts one or more predicate logic expressions and determines whether each of those expressions is true or false. Predicate logic works well as a

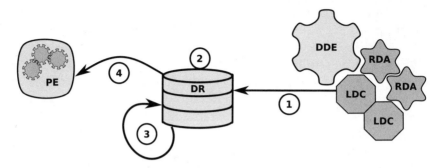

Data Repositories
1. Identify necessary information from local sources.
2. Store information in an efficient, retrievable way.
3. Annotate information ith source and time.
4. Index information for rapid retrieval.

Figure 11.5 DRs collect information from multiple sources over time.

tool to determine whether an assumption is valid: a specific, measurable assumption can be formulated as a predicate logic statement such that if the statement is true the assumption is valid[1]. The fact that predicate functions have a Boolean result does not imply that predicate evaluation is not computationally intensive. Assumptions may require complex expressions to fully capture their validation criteria.

The LCP issues commands to configure applications if certain assumptions associated with the application fail to validate. The LCP implements the training wheels portion of the pattern, in that it takes proactive steps to prevent some error condition from propagating through the application or through the node where the application resides. The LCP issues controls as a result of the actions of the PE with the purpose of impacting one of more applications on the node.

11.3.2 Control and Data Flows

Data flows through this pattern (see Figure 11.6) create subtle dependencies that must be well understood when implementing the PE and assessing the impact of actions taken by the LCP. The act of running commands on a local node may change the state of the local node, and thus, the values reported by the LDCs in the system.

LDCs and RDAs collect information from local and remote data sources, respectively, into the DRs instantiated on a node. Their interaction with the DR is one way, as they only sample the current state of the node (in the case of an LDC) or the reported state of a remote note (in the case of an RDA). The only way to change the values produced by either of these elements is to change the state, and thus, reporting, of the nodes in question. Similarly, the DDE forms a closed loop with the DR in which information from the DR is read, new data items are derived, and then those new items are written back to the DR (or original values are overwritten).

---

1. Alternatively, if the statement is false, the assumption is valid. Either method can be used in the construction of this pattern.

11.3 Pattern Overview

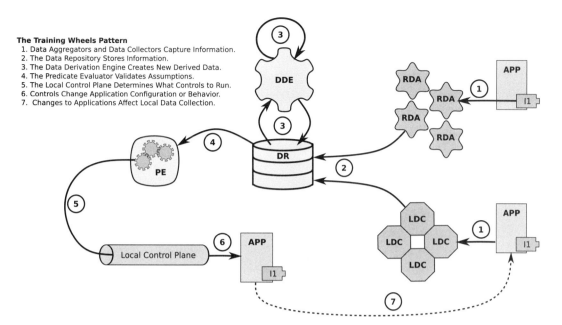

Figure 11.6 Data flows through the TWP.

The PE in the system can be configured with multiple predicates associated with the assumptions needing validation in the system. At periodic intervals, the PE can evaluate predicates by reading their data values from the DR and determining whether any controls need to be run. When necessary, information associated with controls to be run is communicated to the LCP. Notably, the PE does not directly alter the values stored in the DR. The LCP, by running commands on the local node, may change its state, causing new values to be generated by the affected LDCs which will result in new values being placed into the DR.

Like many autonomy patterns, the TWP creates a control loop around certain data items (with the added element of off-node information). While this can be a powerful construct, the coordination of component interactions must be structured to avoid data inconsistencies. For this reason, the flow of control through the system requires more engineering than the flow of data. There are three main phases associated with the evaluation of a predict: data collection, evaluation, and reaction, with each phase being run through for each predicate defined by the system (Figure 11.7). There are two ways to implement these phases: lazy evaluation and strict evaluation.

In a lazy evaluation approach (see Figure 11.8), multiple phases of evaluation may be active at once as part of different applications or threads of control on a node. The evaluation is considered lazy as predicates are collected and evaluated on an as-needed basis. The LCP may be running a command at the same time an LDC is collecting data and a PE is evaluating a predicate. Alternatively, a strict evaluation approach involves performing all steps of one phase prior to all steps of a subsequent phase. In such a strategy, all data necessary to review all rules would be sampled, then predicates analyzed, then local commands run.

There are several processing benefits to a lazy evaluation approach. PEs can evaluate whenever there exists processing time and other resources and will see

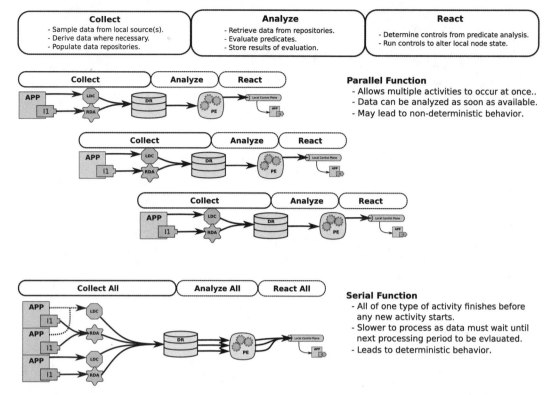

Figure 11.7 Timing data collection to avoid circular dependencies.

efficiencies from being deployed on multiprocessor or multicore systems. The wall clock time measured for performing evaluations should be less as more operations occur at the same time, and this leads to generally faster reaction times. The single—but significant—drawback to a lazy evaluation approach is that it can lead to nondeterminism in how it applies its reactions, and thus, the end state of the node itself. When a reaction to a predicate failure changes a collection value for some other predicate, then there is a dependency between the reaction of the first predicate and the collection of data for the second predicate. If there is ambiguity on how that sampling is ordered, there will be ambiguity on what actions are taken by the system. A strict evaluation approach avoids this issue because all data are collected first, meaning that all predicate evaluations are performed using the same set of collected data. This effectively avoids any ambiguity by guaranteeing a consistent and unchanged set of inputs for the PE.

11.4 Service Types

The mapping of pattern components to service layers (see Table 11.1) is driven by the most logical layers of the service stack to instantiate data repositories.

Any application at any service type can serve as an LDC because any such application may generate data that helps to define local node state. Similarly, any application which has the potential to interact with remote nodes (or to look at data

11.4 Service Types

Consider three sample predicates evaluating an integer data value: "A":

- Predicate 1: IF (A > 3) THEN SET(A = 0)
- Predicate 2: IF (A > 3) THEN SET(A = 1)
- Predicate 3: IF (A > 3) THEN SET(A = 2)

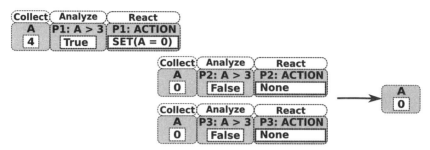

Lazy evaluation
- Predicates 2 and 3 evaluated after the reaction to predicate 1.
- Since the reaction to predicate 1 changes the collected value of "A", predicates 2 and 3 will have different inputs than predicate 1.
- This leads to a final result of A = 0 in this example.

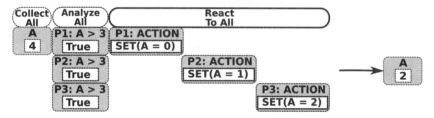

Strict Evaluation
- Data collected once and shared for all predicate evaluations in the system.
- Predicates 1, 2, and 3 all use the same value for "A".
- After all predicates evaluated, then the reactions are all performed according to the ordering defined in the system.
- This leads to a final result of A = 2 in this example.

Figure 11.8 Lazy verrsus strict evaluation of predicates.

from remote nodes) can serve as an RDA. In this sense, data can be generated from any layer in the system.

Not every service layer needs to implement its own DR and attempting to do so unnecessarily complicates the system. There are only two service layers where data need to be centralized: networking and user applications. A DR at the networking layer captures information about the physical network and the physical node in that network. In cases where there is a logical network (implemented as either a federation or an overlay), applications may not have insight into the physical

Table 11.1 Training Wheels Service Types

Service	LDC	RDA	DDE	DR	PE	LCP
I. Transport	X	X				
II. Networking	X	X	X	X	X	X
III. Federating	X	X				
IV. Overlay	X	X				
V. User endpoint	X	X	X	X	X	X

network layer and may not be able to access a DR with physical layer scope. In this case, implementing the DR at the user layer provides allows for the collection of information for logical networking and user applications. Other pattern elements such as the DDE, PE, and LCP all require a DR to operate, and thus, can only be implemented at layers where the DR exists.

11.5 When and How to Integrate

The TWP provides the building blocks for a predicate-based autonomy system that determines whether the current state of a node matches its assumed state and how to handle cases where it does not. This section describes the times when this pattern should and should not be used and how to make design decisions on the mapping of pattern roles to software applications.

11.5.1 When to Use This Pattern

This pattern is useful when the correct operation of applications on a node requires making assumptions on the likely state of the network in situations where communications with other parts of the network may not be available. The pattern implements a deterministic autonomy function that helps applications select from a series of predefined responses to predefined criteria. The TW pattern should be considered when the following conditions are true.

Proper function is predicated on specific, measurable, dynamic assumptions. Specificity and measurability are two characteristics of a verifiable assumption. A specific assumption is one for which the data necessary to determine validity are well defined. A measurable assumption is one for which those data can be unambiguously collected and evaluated. All applications make some assumptions as part of their operation; dynamic assumptions are those that must be checked more than once as their validity may change over time. The TWP captures specific, measurable, dynamic assumptions as predicates that can be analyzed when necessary by one or more applications on a networked node.

Assumptions can be validated on the local node. This pattern provides a framework for assumption validation and review without reliance on services residing elsewhere on the network. High-availability networks such as the terrestrial internet use external services (e.g., DNS, NTP, and CAs) whose round-trip response can be built into application operations. These services may not exist—or may not be readily available—in challenged networking environments. When a node can either self-derive or otherwise cache the information necessary to check assumption validity, this pattern can apply.

There exist useful mitigations to failed assumptions. The TWP provides a mechanism for reacting to circumstances when assumptions are no longer valid. While the pattern models assumptions as predicates in the system, it also models responses to predicate evaluation failures as part of local control. Therefore, the primary function of this pattern is to determine when and how a node can perform self-configuration to correct the operating environment of the node. If no such correction can be applied, this pattern becomes a very computationally intensive way to detect an unrecoverable fault.

Network nodes have enough processing resources. The frequency with which predicate evaluation must occur, and the amount of data that must be collected and analyzed as part of that evaluation, can impose a large processing burden on a networked node. The need to serialize the order in which logical elements communicate drives the frequency of the predicate evaluation. If multiple predicates must be examined within a specific timeframe, and access into the system is serialized, then the last predicate evaluated must still fit within that window. While a powerful tool, self-configuration requires a significant resource investment in the nodes that implement it.

11.5.2 Recommended Design Decisions

The detailed design of components, and the way in which they should be integrating into a functioning network, are specific to the eventual operational environment. However, in generating the detailed designs of these components, multiple recommendations exist for how to make design decisions that result in a scalable, efficient, and operational implementation.

Keep predicate expressions independent. Because each predicate corresponds to an assumption being verified in the system, they should be defined as independently from each other as possible. Ideally, predicates should be evaluable in random order as the evaluation of one predicate would not influence the evaluation of some other predicate. Systems should avoid designs that allows the evaluation of a predicate to have a side effect of changing a data value in the system. For example, predicates of the form (A > (B = C + 10)) would have a side effect of altering the value of B, making the order in which the predicates are evaluated meaningful.

Prioritize local configuration changes. Predicate evaluation should not change the state of the node, and therefore, predicates should be evaluable in any order. The same is not true of the LCP. The LCP changes the state of the node to better maintain the assumptions that applications make for their operating environment. The order in which these changes are made may affect the resultant end state of the node, and therefore, order and priority must be part of the consideration for implementing any control plane. An example of the value of such order is when running controls that switch between options such as selecting between a primary and backup component. Order matters in sequences such as power on the backup component, make the backup component now the active component and vice versa, turn off the backup component.

Serialize and centralize predicate evaluation. Conceptually, assumptions are validated at a single instance in time even though the evaluation in any actual system occurs over a noninstantaneous time period. To prevent data sanity issues associated with spanning evaluations over time, implementations of this pattern should serialize and separate the actions of predicate evaluation, and local control. Evaluating all predicates first, and then collecting the set of controls to be run second prevents the execution of a control from influencing the evaluation of some other predicate. Additionally, to optimize resources on computationally constrained nodes, evaluation engines and LCPs should be centralized in their implementation to avoid synchronization issues.

Prefer sampled and derived data to complex predicates. This pattern defined multiple ways in which data can be introduced into the system: direct sampling, derivation, and ad hoc. Direct sampling occurs at the LDCs and the RDAs in a system. Derivation occurs through DDEs, and ad hoc computation occurs in the construction of predicates as they are evaluated in the PE. Where possible, data values should be defined and calculated as far upstream as possible to avoid processing delays and duplicate calculations. LDCs and RDAs act as the initial data entry for the TWP and should provide as much data as possible to DRs, especially in cases where these data are generated in a computationally efficient way, such as in embedded software or firmware. Data generation through the DDE should be used only when creating data products fused from multiple LDCs and/or RDAs where it would be impossible for any individual LDC or RDA to build the data. DDE data generation is inherently inefficient since it must interface with the DR and build on already-sampled data. Encoding the calculation of a data value into a predicate should always be avoided. In cases where multiple predicates use the same derived value, the value would be calculated for each evaluation. Additionally, when predicate evaluators are built as interpreted engines and not compiled source code (or firmware), the calculation is not as efficient as if the data were calculated by an LDC or DDE. Consider the two predicates: $P1 = (A > (B+C))$ and $P2 = ((A - D) > (B + C))$. In this case the value of $B+C$ would be calculated twice, once when evaluating P1 and once when evaluating P2. Instead, the value should be added to the DDE (e.g., $F = B + C$) and have that value referenced in the predicates.

Consider data timeliness. The data used to assess the validity of an assumption is expected to be current at the time of the evaluation. The timeliness of the data, therefore, must be part of the overall predicate analysis. If the data is recent, then predicate evaluation can occur without issue. However, if the data collected was not recent, then the predicate evaluation may be inconclusive as it could be based off stale data. Implementations of this pattern should provide a way to assess the timeliness of data (e.g., by using timestamps) and to provide a mechanism by which the PE can determine whether stale data may be used in the evaluation.

11.6 What Can Go Wrong

There are very few guarantees in highly challenged networking environments, making the deployment of this pattern susceptible to a variety of potential issues.

Circular dependencies. The activities defined in this pattern conform to a stimulus-response system, in which assumptions, if not valid, result in a mitigation being applied on a local node. However, the application of a mitigation on the node might cause some other assumption to itself be invalidated. It is possible to construct a system that can result in an infinite loop of reactions. When this happens, a mitigation cascade can prevent the system from becoming balanced again. The configuration of mitigations requires some method to prevent this situation, either by applying formal methods or systems engineering to ensure that the system is stable.

Processing time. In situations where the system is invoked periodically (such as every second), then the steps associated with the system (data collection, predicate evaluation, and local control) must finish prior to the start of the next period. In cases where the system does not run on a dedicated processor, there must be enough processing time left after predicate evaluation to perform all other responsibilities on the processor. As the number of data values and predicates in the system increase over time, the overall performance of the system may degrade unless maintenance is undertaken to remove or reformulate older items.

Stale mitigations. Over time, the function of applications on a node may change such that the local controls necessary to successfully implement mitigations need to be updated. When this pattern is used because a node is often disconnected from central authority sources, updating mitigations in a timely fashion may be impossible. This can lead to a situation where a node self-configures itself to invalidate other mitigations configured on the node.

11.7 Pros and Cons

The TWP provides a way for applications to leverage information known locally to a node to verify assumptions that would otherwise require coordination with remote applications across the network. To be applicable, application and their assumptions must have the right set of characteristics. This section summarizes some of the benefits and pitfalls (see Table 11.2) of using this pattern on a networked node.

11.8 Case Study

This section provides an example of this pattern applied to spacecraft fault protection autonomy systems. This is an example of the TWP instantiated at the application service layer where responses to spacecraft fault conditions cannot be rectified in time by waiting for operators in the loop.

11.8.1 Spacecraft Fault Protection

Spacecraft fault protection systems (see Table 11.3) typically enforce dynamic assumptions associated with the proper functioning of onboard spacecraft systems. These assumptions are stated as some measurable indication of health is judged to be within a nominal range and if the assumption fails, then a mitigation is performed in an attempt to bring the system back to nominal operational parameters.

Table 11.2 TWP Pros and Cons

Design Goals	Pros	Cons
Validating assumptions	Making assumptions specific and measurable to represent them as predicates.	Nodes might not have timely information necessary to validate them.
Self-configuration	An LCP runs commands on the node in response to the evaluation of local node time and state.	Tracking dependencies among configurations is difficult. Significant effort must go into keeping the system stable.
Graceful degradation	Only failed validations trigger local changes. These changes can be made specific to the predicates that did not validate, and only for the period of time that they do not validate.	Mitigations must be prioritized so that they do not contradict each other. Care must be taken to ensure data integrity.
Cost and complexity	Restricting local control to handling critical assumptions is simpler than open-ended autonomy systems. Providing a local autonomy function is far less costly and complex than ensuring remote applications are always available.	This pattern requires significant resources on the node to collect data and provide timely evaluation and control execution.

Table 11.3 Examples of Spacecraft Fault Protection Assumptions

Type	Assumption	Mitigation
Component health	The currently selected component is healthy and ready for use.	Change to using a backup version of the component.
Thermal control	The internal temperature of the spacecraft is within a nominal range.	Turn on heaters or open louvers to alter temperature until it is in range.
Commanding configurations	The spacecraft has been commanded within a reasonable amount of time indicating that the communications system is functional.	Degrade into a safe mode and alter communications configurations to reestablish commanding with mission control.

The NASA Goddard Space Flight Center (GSFC) provides an open-source collection of spacecraft flight software called Core Flight System (CFS) [4, 5]. Each service in the CFS is implemented as an individual application, and one such application—the limit checker (LC)—implements a fault-protection autonomy system which can be mapped to the TWP (see Figure 11.9).

The LC application collects data values from other CFS applications over the CFS publish-subscribe messaging bus. Collected data values are used to calculate the values of one or more predefined watch points in the system. Once watch points have been calculated, a series of one or more action points are evaluated; if the predicate associated with an action point is true, then one or more commands are issued by the LC application as new commands into the CFS system (Table 11.4).

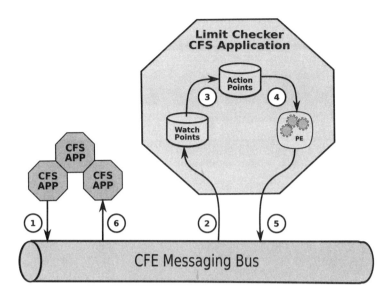

The CFS Limit Checker Data Flow
1. Applications produce data to be used in checking limits.
2. The LC App subscribes to the data it needs.
3. The LC App derives new data called "watchpoints" by combining existing information.
4. Action points define predicates and mitigations. An analyzer determines which action points to run.
5. The LC App published comands to other applications.
6. Other applications run commands sent from the LC App.

Figure 11.9 The CFS LC application conforms to the TWP.

11.9 Summary

All applications are built with some set of assumptions related to their operational environment and the characteristics of their user input. These assumptions are used to make fundamental decisions relating to the design of the application software and the architecture of the network where the applications will operate. Part of the function of an application is to ensure that the environment in which it operates, and the inputs it is given, conform to these assumptions. Terrestrial networks accomplish this task by using centralized network services that can help applications determine whether their assumptions are being met. In DTNs, applications must perform the same function using resources that may only be available on the local node.

The TWP provides a way to build a constrained, deterministic autonomy function on a local node for the purpose of validating assumptions. In cases where the operating environment or user inputs do not conform to these assumptions, this

Table 11.4 Mapping Pattern Components to the LC Applicatiiona

Pattern Component	LC Application Component	Description
LDC RDA	Other CFS Applications	CFS applications generate data that is subscribed to by the LC application and used to calculate watch point values.
DDE	The watch points table	The LC application defines watch points whose value is calculated from the values of collected data values and other watchpoint values.
PE	The action points table and table processing	Action points in the LC application consist of a predicate comprised by watch points that, if true, will result in running of a command or a series of commands.
LCP	The CFS Messaging Bus	When an action in the action points table should be run, the associated command or set of commands is published to the CFS messaging bus for execution by the spacecraft,

pattern provides a mechanism for altering the local state of the node to help resolve discrepancies. While no mechanism can discover the state of remote nodes when the local node loses network connectivity, this pattern does allow for the caching and inspection of remote information as well as the collection of local information. An example of this pattern was provided as part of an open-source spacecraft flight software system used to check functional limits for aspects of flight software and to issue correcting commands when values are out of limits.

11.10 Problems

11.1 Provide an example of a static assumption made by a networked application. Describe why this assumption only needs to be verified once and discuss how this verification should be built into an application.

11.2 Provide a nonsecurity-related example of a dynamic assumption made by a networked application. Explain how often this assumption would need to be validated and what the validation mechanism would be.

11.3 Provide an example of a control plane used on the terrestrial internet. What kind of information is passed on this plane? Would an application continue to work if network connectivity has been lost? Why or why not?

11.4 Discuss how an application would be able to switch between taking direction from a remote control plane and a local control plane in instance where both are resident on a networking node.

11.5 Consider a temperature-sensing node which turn on a heater if the local temperature on the node goes too low. Explain the role of the LDC, DR, PE, and LCP in this situation. What additional functionality would justify the use of both an RDA and a DDE in this example?

11.6 Provide a heuristic by which a DR might know when it is acceptable to remove a data value to save storage space.

11.7 Provide an example where a DR may receive the same data value from multiple sources. Explain what type of logic could be implemented to prevent this situation from negatively affecting the functioning of this pattern.

11.8 Provide an example of a dynamic assumption codified as a predicate.

11.9 Provide an example of a mitigation that can be implemented by an LCP to fix a failed assumption on a node. In your example, explain why the original assumption would fail to validate and why it would succeed in evaluating after the mitigation had been applied.

11.10 Describe a mechanism by which some portions of the TWP could be run in parallel without risking nondeterministic results.

References

[1] Mannion, M., and B. Keepence, "SMART Requirements," *ACM SIGSOFT Software Engineering Notes*, Vol. 20, No. 2, 1995, pp. 42–47.

[2] Kummler, P. S., L. Vernisse, and H. Fromm, "How Good are My Requirements? A New Perspective on the Quality Measurement of Textual Requirements," *2018 11th International Conference on the Quality of Information and Communications Technology (QUATIC)*, IEEE, 2018.

[3] Bjerke, M. B., and R. Renger, "Being Smart about Writing SMART Objectives," *Evaluation and Program Planning*, Vol. 61, 2017, pp. 125–127.

[4] Cudmore, A., "NASA/GSFC's Flight Software Architecture: Core Flight Executive and Core Flight System," *NASA Flight Software Workshop*, 2008.

[5] McComas, D., S. Strege, and J. Wilmot, "core Flight System (cFS): A Low Cost Solution for SmallSats," 2015.

CHAPTER 12
The Stow Away Pattern: Annotated Messaging

This chapter presents a mechanism for applications to exchange data without requiring new messages in a network. The pattern provides a way in which existing messages can be annotated with additional information to reduce the overall number of messages sent by a node. Because every message in a network incurs processing overhead associated with security, routing, management, storage, and transmission time/energy reducing the number of messages reduces the inefficiencies of nodes in a network. This strategy can be used to consolidate all message-related data into a single message or to propagate configurations or network measurements along a given messaging path.

The stow away pattern (SAP) provides a mechanism for network applications to extend the concept of a message payload to include additional annotative information. These payload extensions may exist within a message from the message source to destination, may be inserted or removed during the route, and might be updated by waypoint nodes. To accomplish this behavior, the SAP relies on either transport protocols supporting secondary message headers or conventions for building structured payloads. Just as the term stow away refers to a person who attempts to travel without paying a fare, these payload extensions ferry information without incurring the overhead cost of individual messages in a network.

12.1 Pattern Context

Networks conventionally support a one-to-many mapping between applications and network messages: a single application generates multiple payloads which are encoded into messages and sent over a network to some peer application which ultimately receives those payloads.

When discussing the relationship between applications and networked communications, the terms payload and message cannot be used interchangeably. For example, application developers may colloquially refer to the message that is generated by the application, but the material produced by an individual application is not a complete network message. Application data must be annotated with additional information necessary to properly address, secure, and route the data and the application data itself may require multiple encodings to be transported by the network. Most network messaging protocols (see Figure 12.1) define a common

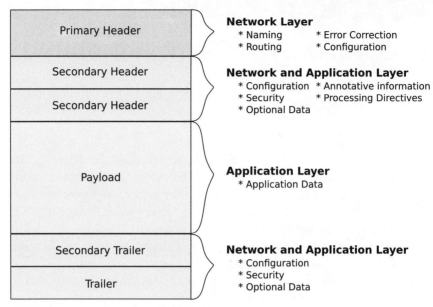

Figure 12.1 Common message protocol structures.

logical structure and nomenclature to differentiate application data from other types of annotative data required by the network.

A header is a preamble that includes information about how the message is encoded (such as by specifying a protocol version or other magic number) and logical information necessary for processing (routing, storing, validating) the message. Messages may have more than one header, in which case there is typically a concept of a single, mandatory primary header and one or more optional secondary headers. The primary header ensures that message protocol agents can rely on a specific set of information with a specific encoding to begin coherent message processing. Secondary headers can specify additional, optional information as identified by options set in the primary header. There is typically a single payload in a message, which consists of encoded application data and any additional annotations (such as byte length) necessary to process that data. Finally, some message protocols also support the concept of one or more optional trailers, which follow the payload.

The decision to place information before the payload (header) or after the payload (trailer) depends on assumptions relating to the size and character of the payload [1]. For relatively large payloads, it may be advantageous to place information in a trailer so that a processing node does not need to store header information while processing the payload. For example, some message protocols include an asserted checksum to ensure nothing was corrupted during transmission. A receiving node can calculate an actual checksum over the received message, compare it to the asserted checksum, and infer that the transmission was corrupted if the checksums do not match. Placing the asserted checksum in a header would require a processing node to store it while calculating the actual checksum as part of message processing [2]. While a seemingly small amount of storage, it can grow quickly when messages are received at a very high rate. Placing the asserted checksum in a trailer allows it to be read and used in real time without incurring extra storage overhead.

The overhead associated with a message refers to the amount of nonapplication data that must be included in the message to account for necessary encodings and annotations. This value can either be computed as an absolute measure (such as number of bytes) or as a relative measure (such as the percentage of overall message size) (Figure 12.2). Absolute measurements are useful for predicting the amount of memory and processing necessary to handle messages through the network and sizing buffers and queues in software and hardware. Relative measurements provide an indication of the overall efficiency of the network.

One way to reduce the amount of absolute overhead in a message is to synchronize state information at the endpoints of the message exchange and then include references to that information in the message itself. If a message can assume preexisting information at its source and destination, then there is significantly less information that would need to be carried by the message itself.

One example of this approach to overhead reduction is the security association used by the IPSec [3] protocol. As part of establishing a secure communications channel, endpoints must exchange information relating to how security services will be configured and used. Once established (or preplaced) these associations can be referenced by IPSec messages through a simple security association identifier (SAID). Because each secured message only carried the SAID, the messages are smaller than if they carried the entire set of negotiated/configured parameters. Conversely, if the endpoints of the exchange lose synchronization, the SAID may become meaningless and cause messages to fail to authenticate or decrypt.

Encoding endpoint references in messages can be effective in networks that support end-to-end information synchronization (Figure 12.3). They are not effec-

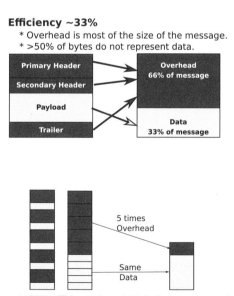
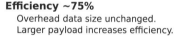
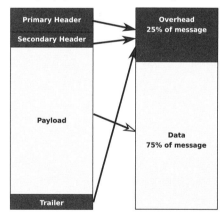

Figure 12.2 Absolute and relative overhead measurements.

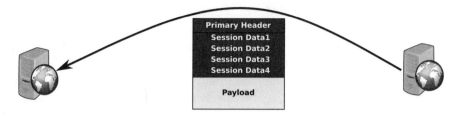

Without shared state at message endpoints, individual messages must carry important annotative information.

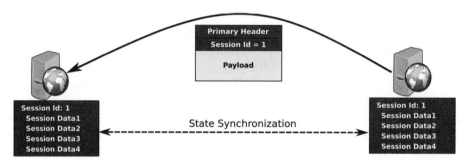

With shared state at message endpoints, individual messages can reference this information to reduce message overhead.

Figure 12.3 Endpoint references reduce message overhead.

tive in chalenged networks that cannot guarantee synchronization, timely message exchange, or even that endpoints coexist simultaneously in the network.

The SAP provides a method for reducing messaging overhead without relying on end-to-end state synchronization by finding ways to intelligently aggregate application information into a single message. Secondary application data can stow away on existing messages for distribution in the network.

12.2 The Problem Being Solved

Nodes communicate both control and data plane information across a network. The rate and volume of information over the data plane is governed by the number of applications that are resident on the node and the amount of data they wish to communicate. The rate and volume of control plane information is governed by the number of network services running on the node and how often they need to synchronize or otherwise report their status information across the network. Regardless of how and how often information is produced, it is the ability of the underlying network that determines the success or failure of message transmission and delivery. Reviewing the ways in which links fail to delivery messages in a challenged network provides insight into alternative ways of structuring messages to increase the probability of successful delivery.

Beyond the use of store-and-forward protocols (such as BP), applications and network services can increase the overall efficiency associated with messaging as a way of improving the likelihood of successful message delivery. Message efficiency in this context refers to increasing the goodput over the link; in a perfectly efficient messaging exchange, the goodput of the link would be equivalent to the throughput of the link.

There are two conditions which negatively impact the goodput of a link: overhead and retransmission. Retransmission decreases goodput because the same application data is communicated over a link more than once. Overhead decreases goodput because, by definition, overhead is the addition of nonapplication data to a message (to include messages with no application data at all).

Traditional approaches to reducing retransmissions include building highly reliable links [4], increasing link throughput, and fragmenting messages [5, 6]. Reliable links reduce retransmissions by lessening the likelihood that a message would be corrupted in transit and, therefore, need to be retransmitted. Increasing link throughput reduces the impact of retransmissions; if a link has a low utilization relative to the data volume it must transmit, then retransmissions have a much smaller impact in the system. Message fragmentation reduces retransmission by allowing portions of a message to be received such that if a portion of a message is corrupted, only a subset of the message would need to be retransmitted, rather than the entire message.

Each of these approaches can be difficult to implement in a challenged networking environment. Where nodes are power constrained, increasing data rates may be impractical. In cases where higher data rates can be achieved, periods of disconnect from the network may have resulted in a large backlog of messaging making it difficult or impossible to achieve the low link utilizations necessary to accommodate retransmissions without messaging impact. Finally, fragmentation does help reduce retransmissions, but at the cost of decreasing message efficiency, as each message fragment further requires its own set of messaging headers, trailers, and encodings.

In cases where increasing link quality to eliminate retransmissions is impractical, the only other approach to increased goodput in a system is to reduce message overhead. The less overhead that appears in a message, the more application data that can be communicated in each data volume from a node. Traditional approaches to reducing messaging overhead depend on whether absolute or relative overhead is being reduced.

Reducing absolute overhead involves reducing the number of nonapplication data bytes in a message by reducing and/or compressing information in each message. Reducing information can be accomplished by using different protocols that require less information to function, or by configuring existing protocols to transmit a smaller volume of annotative information. For example, the IPSec protocol uses a SAID in lieu of a verbose security context to remove a large amount of information from individual IPSec messages. Compressing information can also be accomplished by using more efficient encodings. For example, messages which use ASCII encodings such as XML result in significantly larger message sizes than binary encodings, such as using the concise binary object representation (CBOR) [7].

Reducing relative overhead involves adjusting the ratio of overhead-to-application data in a given user message. This can be accomplished either by reducing the absolute overhead within a single message or increasing the size of the application data stored in a message. In cases where the application data size is large relative to messaging overhead, there may be little gain to optimizing absolute overhead associated with a message.

While absolute overhead may help analyze a single message, relative overhead better quantifies the achievable goodput over a link. For example, if a message originally required 100 bytes of overhead but, through redesign, the overhead is reduced to 50 bytes, then the absolute overhead of the message is smaller by 50%. However, the relative overhead, as a function of the overall message size, is driven by the amount of application data in the message. In this example, if the application data size were 100 bytes, the relative overhead would have been reduced from 100% down to 50%. However, if the application data size were 10,000 bytes then the relative overhead would have gone from 1% to 0.5%—a much less impactful reduction from a networking perspective.

High-availability, high-rate networks over-resource their links so that their link utilizations do not reach high percentages. This allows for message inefficiencies to exist in the network; if there is plenty of throughput, the goodput can be lower and still get messages delivered. This same approach does not scale to challenged networks, where over-resourcing links may not be possible and where relative overhead can become a very large percentage of overall message size because information cannot be kept at messaging endpoints. A mechanism is needed which allows challenged networks to carry annotative information in a way that minimizes messaging overhead and optimizes goodput through the node.

12.3 Pattern Overview

The SAP provides a structure and set of behaviors for augmenting existing messages in a system to carry additional information in a challenged network. By stowing away on an existing message, additional information can be injected into the network without requiring the creation of a new message and the overhead associated with that message. By recoupling the stow away from the message itself, the additional information does not necessarily need to be retransmitted if a message is lost. The pattern is comprised of three logical components: Extensible Payload System (EPS), Payload Extension Processor (PEP), and an Extension Cache (EC) (see Figure 12.4).

12.3.1 Role Definitions

An extensible payload system (EPS) is one that allows different data sets to co-exist within a single message and remain differentiable from one another. The EPS accepts multiple data sets (typically a primary message payload and a series of one or more additional, secondary information) and determines how these sets are best represented in a messaging format. There are two common ways in which payload extensibility can be implemented by an EPS: secondary headers or a structured payload.

Secondary headers refer to the practice of building multiple structures (headers) within a message such that each header can be individually addressed and analyzed. The use of secondary headers requires the use of a protocol which supports them. An example of using secondary headers is the BP, which formalizes extension

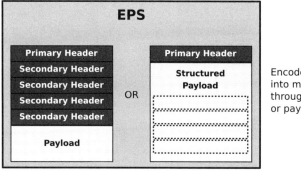
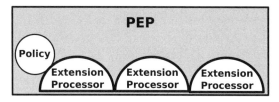
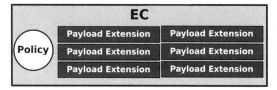

Figure 12.4 System roles of the SAP.

blocks that serve the purpose of secondary headers within bundles. When secondary headers are not available, a structured payload can be used instead. In this approach, a payload can be redefined to allow for individually addressed information. This is like the protocol concept of encapsulation in which a structured message is encapsulated as the payload of an encapsulating message. Examples of protocols that create structured payloads are the Cryptographic Message Syntax (CMS) [8] and the Bundle-in-Bundle Encapsulation Service (BIBE) [9], each of which creates a single envelope payload holding structured data.

A payload extension processor (PEP) refers to software that must exist at waypoints and destination nodes that can meaningfully process payload extensions. This involves determining whether the current node should add a new payload extension, whether an existing extension should be processed and possibly updated by the current node, or whether an existing extension should be removed from a message. In cases where extensions are houses in secondary headers, this is a relatively straightforward operation of adding, removing, and updating header information. In cases where a structured payload is used, then the payload must be read, parsed, updated, and then replaced in a message.

An extension cache (EC) refers to storage of payload extensions when they are not being transported in a message. This is the case when extensions are first being constructed on the node that adds them into a message, when extensions are read out of a message for modification, or when extensions are removed from one message and the stored to be added to a different message at a later time.

12.3.2 Control and Data Flows

The data flows through this pattern are relatively simple, with the complexity being an understanding of when a payload extension should or should not be added to a message. However, this business logic decision is based on the type of information being added and is not considered part of the mechanism itself.

There are three modes governing the data flows through the SAP: inserting, processing/updating, and removing payload extensions. When a node determines that a payload extension should be added to a message, the extension is taken from the EC by the PEP and then encoded in the EPS. As part of processing a received message, a node may use the PEP to read every payload extension in the message, and as a function of the policy of the PEP, the extension may be copied into the local EC, updated, and/or removed from the message. The decision to remove a payload extension does not imply that the payload extension is no longer valuable in the network. It simply implies that the current message is no longer the appropriate bearer for that extension.

The control flows through the system for the SAP are centered on interactions with, and decisions made by, the PEP. In this mechanism, the PEP analyzes local node state, configured policy, and the extensions found in a message and determines from this information how to process each message. There are several decisions that a PEP must make as a function of processing an extension, including the following:

- How to process an unknown type of payload extension;
- How to determine that a payload extension should be removed from a message;
- Whether a payload extension should by copied from or moved from the EC into the EPS;
- How long an extension should stay in the EC;
- Whether there are dependencies between extensions found in a message;
- Whether an existing extension in a message should be updated or whether a new extension should be added to a message.

The policy associated with these decisions may be made globally for a network, may be unique for each PEP on a node, or may be unique for each type of extension in a message. The level of required fidelity will be driven by the overall computational resources of the node, differences among payload extensions, and how frequently payload extensions are included in a message.

12.4 Service Types

The mapping of pattern components to service layers is driven by the most logical layers of the service stack to instantiate data repositories.

The PEP and EC can exist at any service layer that also instantiates the EPS, because the PEP and EC ultimately work together to alter the contents of the EPS. Therefore, the service layers associated with the SAP can be defined entirely by where it makes sense to instantiate an EPS (Table 12.1). The transport, networking, and user service tiers provide the best places to implement the EPS. Federating services broker between transport layers and overlays convert between transport layers making any implementation of an EPS in their service tier redundant to implementing the EPS at the transport tier.

In cases where the EPS uses the existing capabilities of the underlying transport layer, then the EPS responsibilities can be implemented as a transport service mapping each payload extension to a transport messaging structure. In cases where there is no such capability the payload itself must be structured. This payload structuring can be performed as a network service or in user space (Figure 12.5).

When performed as a network service, the user payload is combined with one or more network-layer extensions in a way that is hidden from user applications. This allows the network to augment its operation without requiring a supporting transport protocol and without requiring every user application to support structured payloads. The risk of this approach is that the payload in this case has been effective encoded by the networking layer and must be decoded later before the payload can be returned to a receiving user application.

When performed as a user service, structuring a payload occurs prior to the payload being added to the networking layer. This case provides users the most control over how extensions are added in the system, but also required that user applications accept responsibility for defining and processing extensions. In this scenario, a structured payload cannot add, alter, or remove payload extensions at waypoint nodes because only the user endpoints understand the structured payload.

12.5 When and How to Integrate

The SAP provides a mechanism for adding payload extensions to a message to reduce the overall number of messages in a system and/or to ensure that a series of related information arrives together (Figure 12.6). This section describes the times when this pattern should and should not be used and how to make design decisions on the mapping of pattern roles to software applications.

12.5.1 When to Use This Pattern

The concept of stow away information in a message is not widely used in the terrestrial internet because its high-availability and high data rates do not experience the problems that this pattern solves. However, this pattern can be powerful and

Table 12.1 Stow Away Service Types

Service	EPS	PEP	EC
I. Transport	X	X	X
II. Networking	X	X	X
III. Federating			
IV. Overlay			
V. User endpoint	X	X	X

Transport Layer Structured Headers

Certain transport layer protocols provide a natural extensibility mechanism through the use of "secondary headers" and "trailers". This allows new information to be added to a message without changing the size, processing, or encoding of the message payload.

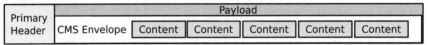

The Cryptographic Messaging Syntax

Encodings such as the CMS provide a hierarchical structure where information can be nested. One nesting structure in the CMS (the envelope) allows for different types of content to be placed inside of it. Using a standard such as CMS allows applications and other nodes to add and remove information from the CMS payload without changing the ability of other nodes or applications to access their own information.

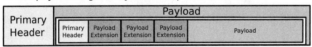

The Bundle In Bundle Encapsulation Service

The BIBE service is proposed as an encoding mechanism for the Bundle Protocol in which a structured bundle can be used as the payload of another bundle. In this case, the encapsulating bundle can be considered to have a structured payload. The structure of this payload is given by the encapsulated bundle.

Figure 12.5 Different ways of structuring an EPS.

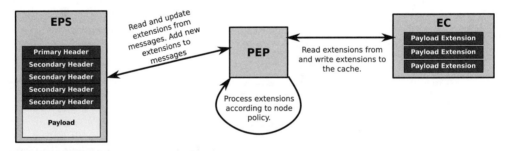

Figure 12.6 Data flows through the SAP.

useful when the quality of links in a challenged network makes the likelihood of message delivery (and in-order message delivery) difficult. Cases where coordinated messaging and overall goodput in the network are not guaranteed benefit from this pattern. More specifically, the SAP should be considered when the following conditions are true.

Session state cannot be synchronized at message endpoints. A large amount of terrestrial internet traffic occurs in the context of sessions, where endpoints pre-negotiate network information. This information can be assumed to exist by individual messages, and therefore, messages can omit this information from their own headers and payloads. In cases where this is not the case, messages must carry this information with them creating a situation where a message must contain both user and network data. The SAP can provide a mechanism for ensuring that every message in a challenged network can stand alone and be fully interpreted on its own.

Annotative payload information exists that must be treated separately from the payload. In cases where sessions are not used for information exchange, protocols over the terrestrial internet may rely on guaranteed, in-order delivery of messages to send annotative information to a destination first, followed by user information second. In cases where in-order delivery of messages cannot be guaranteed, this kind of annotative information can be added to the message itself.

Stow away data share destinations and are small relative to message overhead. Every network includes some amount of control data associated with configuring and maintaining network services such as security, routing, and management. In cases where these messages are small relative to per-message overhead there may be an efficiency in attaching this information to an existing message that shares the same destination. By merging information that is destined to the same node the goodput of the network could be increased for otherwise relatively little effort.

Stow away data informs message paths and not user endpoints. Control plane information can be useful not just to the endpoints of a user message, but also to every node along the messaging path. For example, a node may wish to communicate its congestion predictions to every downstream node along a message path in which case that information could be appended to a message traversing that path. Alternatively, network operators may wish to collect and report on information collected at every node visited by a message along its path, in which case a payload extension can be placed on the message that is updated at every hop along the path.

12.5.2 Recommended Design Decisions

The detailed design of components, and the way in which they should be integrating into a functioning network, are specific to the eventual operational environment. However, in generating the detailed designs of these components, multiple recommendations exist for how to make design decisions that result in a scalable, efficient, and operational implementation.

The EC should not be long-term storage. Information stored in the EC should not be considered as a replacement for other persistent storage on a node. In cases where payload extensions are used to place session-like information in a message body, this information should be stored in the EC no longer than the natural lifetime of the virtual session. In cases where payload extensions are used to convey local node state for control plane messaging, these extensions should frequently expire out of the EC and be refreshed with up-to-date information.

Stow away data should have an authentication mechanism. Because nodes along a messaging path may add new payload extensions (or update existing extensions) it is important that some type of authentication mechanisms exists so that PEPs can ensure that they only process valid information. This authentication must be performed on a payload extension by payload extension basis. Other security services, such as encrypting payload extensions may also be useful depending on the type of information and the requirements of the network. In cases where encrypting cipher suites generate their own authentication mechanisms, additional authentication of the payload extension is not necessary.

Uniquely identify every type and instance of stow away data. When payload extensions are added to a message, either in their own secondary headers or as part of a structured payload, they must be uniquely identifiable. PEPs must be able to determine the type of data when processing and this includes cases where there exist multiple instances of the same type of data in a message.

Differentiate types of stow away data for processing. The SAP can be used to add message-related and message-agnostic information to a message. Message-related information includes data that annotate and assist with the interpretation of the message's user payload. Message-agnostic information often is used to implement network control plane functions along the message paths that happen to be used by a message. PEPs should treat these types of information differently when processing these extensions. Message-related information should be kept unaltered with the message, and message-agnostic information may be updated and removed as necessary as a function of the network needs for this information.

Include time information in extensions. Transport layer protocols in a challenged network, such as the BP, use store-and-forward techniques to increase the probability of delivery in cases where network links come and go over time. This adds a time-variance to all network control algorithms and user expectations that is not always present on the terrestrial internet. Stow away data should include time information so that PEPs can determine whether the data continues to be relevant both at waypoint nodes and the destination node.

12.6 What Can Go Wrong

There are very few guarantees in highly challenged networks, making the deployment of this pattern susceptible to a variety of potential issues.

Message fragmentation. Message fragments can alter the proper functioning of the PEP at waypoint nodes in a network. In cases where payload extensions are encoded in a single structured payload, and the payload has been fragmented, any attempt by a PEP to alter the size of the structured payload could prevent message reassembly as the message destination. In cases where payload extensions exist in their own headers, if one message fragment has some headers and another message fragment includes different headers, it becomes impossible for a PEP to determine whether or not a full set of information exists in the message as a whole. This can result in PEP processing errors, particularly when there may exist dependencies among payload extensions in a message.

Routing loops. When payload extensions are used to provide control plane information to nodes along an existing message path topological changes in the network may result in routing loops, both for messages and the extensions that they carry. However, this problem can be exacerbated when a PEP decided to remove control plane information from one message and attach it to another message, presumably to inform a different messaging path. Attempting to creating a routing plane for payload extensions and a separate routing plane for messages is tremendously risky and a discouraged use of extensions.

Increased retransmission penalties. Adding payload extensions to a message increases the overall goodput of the network—in the nominal case—because these extensions do not need their own message and per-message overhead to exist in the network. However, in cases where a message carries with it a significant number of extensions, and that message must be retransmitted, then the overall impact of the retransmission is increased by both the size of the original message and the size of all of its extensions. Particularly over lossy links, the use of extensions to a message should be kept to a minimum.

12.7 Pros and Cons

The SAP provides a way to augment existing message in a challenged network, both to annotate use payloads and to assert control information along messaging paths in the system. This section summarizes some of the benefits and pitfalls of using this pattern on a networked node (see Table 12.2).

12.8 Case Studies

This section provides two examples of this pattern applied to message security and synchronizing routing information.

12.8.1 The BP Security Extensions

The BP Security Specification (BPSec) [9] provides security services for blocks within a bundle. Unlike other security protocols, such as IPSec, which secure a message

Table 12.2 SAP Pros and Cons

Design Goals	*Pros*	*Cons*
Reduce message overhead	Fewer messages (and message headers) reduce both absolute and relative overhead measurements.	Repeating payload extensions in multiple user messages can increase overall overhead.
Reduce retransmissions	Fewer messages means fewer number of message retransmissions.	Individual messages become larger, so retransmissions are more impactful.
Make messages independent of sessions	Messages can now hold session and other annotative information. There is not a need to store this information at endpoints.	Annotative information can expire in a store-and-forward network faster than the payload expires.
configure data along message paths	Extensions can be processed and used independently of user payloads along message paths.	Fragmentation prevents nodes from using extensions at waypoints.

in its entirety, the BPSec enables different integrity and confidentiality services for different parts of a message. BPSec defines two type of extension blocks, the block integrity block (BIB) and the block confidentiality block (BCB), which can be added to a bundle to define these services.

BIBs provide plain text integrity services that target other blocks within a bundle. The contents of a BIB include information about the integrity service used to generate an integrity signature, the signature itself, and the other block(s) in the bundle used to generate that signature. For example, a BIB could be used to calculate a signature for the bundle's primary block, payload block, or any other extension block in the bundle.

BCBs provide a signed confidentiality service that encrypts other blocks in a bundle. The contents of a BCB include information about the confidentiality service—to include key and cipher suite information—as well as the target block(s) in the bundle that have been encrypted. The target block of a BCB will have its plain text content replaced by cipher text generated in accordance with the cipher suite defined by the BCB.

Consider the illustration of a bundle provided in Figure 12.7, which contains four blocks: a primary block (B1), a payload block (B6), and two extension blocks (B4, B5). Using the BPSec specification, a node could choose to provide integrity signatures for the primary block, one of the extension blocks (B5), and the bundle payload. If these three signatures all used the same cipher suite with the same key information, they would be represented in a single BIB block that could be added to the bundle as block B2. A node could also choose to encrypt the other extension block, in which case a BCB would be added to the bundle as block B3 and the plain text contents of extension block B4 would be replaced with cipher text.

In a BP network, waypoint nodes may choose to add or remove extension blocks as part of normal bundle processing. This includes waypoint nodes that choose to add or remove BIB and BCB blocks or add other extension blocks that need their own BIB and BCB blocks. A common architecture where the security

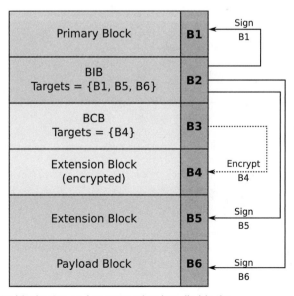

Figure 12.7 Security blocks sign and encrypt other bundle blocks.

associated with a bundle might change at a place other than the bundle originator is when dealing with gateway nodes. A gateway node is one that exists between two or more distinct portions of a network. This can include sections of the network under different administrative control or—in the case of federated networks—different transport protocols. Messages within a specific trusted boundary may only require integrity signing to guard against message corruption, whereas message crossing a boundary may require confidentiality. In that situation, gateway nodes—not the node originating a message—would add BCBs to a bundle and process/remove BCBs at another gateway node.

The design of BPSec follows the mechanism of the SAP as follows. The BP itself implements the EPS role as it provides for the definition of secondary headers, in the form of protocol-supported extension blocks. Extension blocks have a type and a unique identifier and blocks can be processed independently of each other. The BPSec agent running on a bundle node serves as a PEP determining whether and when to add new BIB and BCB blocks into the bundle, whether to process these blocks (such as performing integrity validation or decryption) and whether BIBs or BCB should be removed from a bundle (after they have been processed). The BPSec agent implements a short-lived EC used as a place to store BIBs and BCBs while they are being constructed, but it cannot reuse BIB and BCB contents as they contain signatures and other information specific to the bundle in which they are inserted (Figure 12.8).

12.8.2 Contact Graph Routing Extensions

Contact Graph Routing (CGR) [10] refers to a routing strategy whereby each node in a network is configured with a routing table that includes start and stop times for individual contacts. Routing tables on the terrestrial internet represent instantaneous communications opportunities: If an element appears in a routing table, then the node can send items to those listed destinations. In a challenged network, the CGR lists contacts and when those contacts would be available. This allows a CGR node to determine whether a route can exist at some point in the future and how long a message would need to be stored on the local node prior to being able to take advantage of that route.

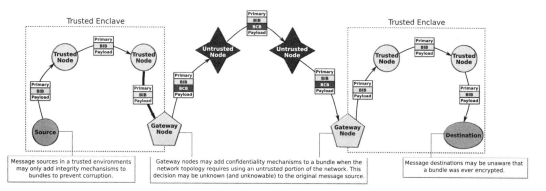

Figure 12.8 BPSec can be added by nodes other than the bundle source.

If each node in a network contains a contact graph (CG) a mechanism must be used to keep each local CG updated. This can be relatively straightforward in cases where data sources and sinks are well-known, regular, and unlikely to encounter congestion. In cases where traffic is more dynamic, an alternate approach is to add CG information to messages traversing the network so that information relating to node congestion can be reinforced along popular messaging paths. In BP networks, this can be used as a CGR extension block (CGR-EB) [11] (Table 12.3).

12.9 Summary

The one-to-many mapping of applications to network messages presumes that the goodput in the network is enough for all necessary messages to be communicated. When the network is unable to effectively communicate messages then the efficiency of the messaging system must be examined. Typical approaches to increasing message efficiency on the terrestrial internet involve improve the quality and capability of network links, but improving links is not practical in a challenged networking scenario. Other mechanisms need to be used to improve the goodput of the network.

Two ways to improve goodput is to reduce the number of retransmissions in a network and to reduce the relative overhead associated with messages in the network. Because reducing retransmissions often requires making links more reliable, and link improvements are not practical in challenged networks, more focus needs to be placed on improving messaging efficiency by reducing overhead.

The SAP provides a way to reduce messaging overhead by provide a many-to-many mapping of applications to network messages. Just as a single application can generate multiple messages, SAP allows multiple applications to combine their information into a single message. By reducing the overall number of messages in a system the overall number of message overhead bytes are reduced. By making data payloads in message larger, the relative overhead of messaging in the system is also reduced.

SAP can be implemented using either transport protocol support for extra message headers or by using mechanisms to create structured payloads. In either case, the type of information added to messages can be message-specific and stay with the message until its destination or message agnostic and used to inform nodes along a messaging path.

Table 12.3 Proposed Contents of a CGR-EB

Attribute	Description
ND	Network distance of encoded path.
Contact[i].start	The time at which the ith contact in the route starts.
Contact[i].end	The time at which the ith contact in the route ends.
Contact[i].resCap	The estimated residual capacity of the ith contact.
Contact[i].rate	The estimated data rate of the ith contact.
Contact[i].rxNode	The address of the node receiving messages transmitted over this contact.

12.10 Problems

12.1 How would you determine whether information should go into a message header or trailer?

12.2 Presume that an integrity signature of a very large payload is included in a message header. Provide a bad-day scenario where having this information in the header and not in a trailer could cause a node to lose processing resources.

12.3 Explain why an absolute measure of per-message overhead can lead to misleading results associated with overall goodput in a network. Defend this explanation with an example.

12.4 Research and provide two items of information that are synchronized at the endpoints of a TCP session. What happens when this information is no longer the same at each endpoint?

12.5 Describe three approaches used in nonchallenged networks to reduce retransmissions. Explain why these approaches might not work in a challenged networking environment.

12.6 Define message efficiency and explain why networks work better with more efficient messages. Explain why goodput is a better measure of efficiency than throughput.

12.7 Provide an example of why a PEP would choose to add the same payload extension to multiple messages going through a network.

12.8 Provide an example of why a PEP would choose to remove a payload extension from a message when the PEP's node does not represent the message destination.

12.9 Provide an example of a payload extension that would stay in an EC for longer than it takes to build the extension and place it in a single message. In this example, explain how the EC would know to eventually expire the extension from its cache.

12.10 Explain a problem that could occur is payload extensions were not authenticated in a network.

References

[1] Stone, J., et. al., "Performance of Checksums and CRCs over Real Data," *IEEE/ACM Transactions on Networking*, Vol. 6, No. 5, 1998, pp. 529–543.

[2] La Porta, T. F., and M. Schwartz, "Architectures, Features, and Implementation of High-Speed Transport Protocols," *IEEE Network*, Vol. 5, No. 3, 1991, pp. 14–22.

[3] Oppliger, R., "Security at the Internet Layer," *Computer*, Vol. 31, No. 9, 1998, 43–47.

[4] Celandroni, N., and F. Potortì, "Maximizing Single Connection TCP Goodput by Trading Bandwidth for BER," *International Journal of Communication Systems*, Vol. 16, No. 1, 2003, pp. 63–79.

[5] Chang, Y., C. P. Lee, and J. A. Copeland, "Goodput Optimization in CSMA/CA Wireless Networks," *2007 Fourth International Conference on Broadband Communications, Networks and Systems (BROADNETS'07)*, IEEE, 2007.

[6] Chang, Y., et. al., "Dynamic Optimal Fragmentation for Goodput Enhancement in WLANs," *2007 3rd International Conference on Testbeds and Research Infrastructure for the Development of Networks and Communities*, IEEE, 2007.

[7] Bormann, C., and P. Hoffman, "Concise Binary Object Representation (CBOR)," RFC7049, October 2013.

[8] Housley, R., "Protecting Multiple Contents with the Cryptographic Message Syntax (CMS)," RFC4073, May 2005.

[9] da Silva, A. P., S. Burleigh, and K. Obraczka, (eds.), *Delay and Disruption Tolerant Networks: Interplanetary and Earth-Bound--Architecture, Protocols, and Applications*, CRC Press, 2018.

[10] Dhara, S., et. al., "CGR-BF: An Efficient Contact Utilization Scheme for Predictable Deep Space Delay Tolerant Network," *Acta Astronautica*, Vol. 151, 2018, pp. 401–411.

[11] Birrane, E. J., "Improving Graph-Based Overlay Routing in Delay Tolerant Networks," *2011 IFIP Wireless Days (WD)*, IEEE, 2011.

CHAPTER 13

The Network Watchdog Pattern: Distributed Error Detection and Recovery

This chapter presents a pattern for the early detection of, and corrections for, faults in a DTN. The pattern provides a way by which resourced nodes in a network may be identified as network watchdogs with an area of influence, acting as a virtual network operator collecting information and applying coordinated responses. Because DTNs are unable to provide timely information to network operators, network watchdogs can be placed strategically to collect network traffic and apply decisions where necessary.

The network watchdog pattern (NWP) implements a two-tiered watchdog concept by which faults are detected at local nodes distributed throughout an area of influence in a DTN. When one or more nodes detect network-level faults, those detections are passed on the network-level watchdog, which initiates corrective actions as necessary. The concept is taken from the use of software and hardware watchdogs in embedded systems to reset computers unless they proactively and periodically assert their proper, healthy function.

13.1 Pattern Context

Embedded systems are those whose processing components are tightly coupled with their environment. Tight coupling, in this instance, means sharing resources, being spatially colocated, and often integrating into a single housing or enclosure. For example, multiple processing components can be packaged on a single board and multiple such boards can be integrated into a single powered enclosure and deployed into a hard-to-access environment. To maintain the proper operation of individual components, fault protection mechanisms must be implemented on the device. One such mechanism is the TWP, which describes how components apply self-configurations based on local state changes. Fault-protection approaches like TWP allow devices to rapidly react to changing local conditions without requiring the delays associated with an operator-in-the-loop response. The limitation of these mechanisms is that they are unable to provide a system-level view of overall health and coordinated corrective actions.

A system-level view of health is needed when multiple components work together to accomplish some function. These multiple components can be spatially distributed (nodes on a network) or colocated (individual software processes;

computers on a dedicated local network). Consider an environmental measurement system implemented as a series of spatially distributed nodes on a network. The health of the system is defined by its ability to coordinate multiphenomenological sensor measurements, fuse them into data sets, and then mine the data to create weather predictions. Similarly, consider a spacecraft implemented as a series of flight computers on a dedicated local network. These computers run commands, produce data, operate instruments, and navigate the spacecraft. The overall health of the spacecraft is measured across all components and must be assessed as an overall system. For example, if a single component fails and a backup component is used instead, then the system-level view of the spacecraft health may not be significantly affected.

Regardless of how components are distributed, building a system-level view of health (and applying corrective actions to maintain that health) requires the development of three capabilities. First, component faults must be detected and those fault conditions that impact system-level health must be communicated to system-level fault protection mechanisms. Second, a system-level corrective action must be determined. Third, corrective actions, once determined, must be applied to the proper components in the proper order at the proper time.

Of these three capabilities, the most critical is the ability to detect and communicate fault conditions that impact system-level health. Delays in detecting component faults increase the likelihood that the fault can cascade into other faults on the component and across the system. While formulation and application of corrective actions can require significant up-front system engineering analysis, none of that work can be applied unless the fault is detected and communicated with enough time to apply that analysis. A fault in a power distribution unit may endanger the physical safety of the devices it powers. A fault in a heater may cause components to experience physical damage by incorrectly regulating the thermal environment. There are two approaches to fault detection in a distributed system: positive and negative acknowledgements.

Positive acknowledgement (ACK) schemes require that the nominal operation of the system is continuously asserted by components such that any interruption to the periodic receipt of an ACK can be interpreted as an error condition. This scheme is effective at catching error conditions on a component—a component is only able to produce the necessary ACK if it is functioning without error. If the component becomes unresponsive for reasons such as infinite looping or processor reset, the ACKs stop. The limitation of this scheme is that ACKs are generated continuously whenever a component is in a nonerror state, and components spend most of the operational time in a nonerror state. Therefore, this generates a large amount of communication traffic and the processing associated with the generation and consumption of this traffic.

Negative acknowledgement (NACK) schemes only require that components communicate when an error has been identified. In the absence of communication, the component should be in a nonerror condition or recovering from a local error condition that does not require additional corrective action. The benefit of this scheme is that it only generates traffic and incurs processing when the component experiences a system-effecting error. Because such errors are presumed to be infrequent, NACK schemes generate less traffic and require less processing than ACK schemes. NACKs allow alternate approaches to solve networking problems, such as reliable multicast [1]. One limitation of this scheme is that it can only operate in conditions where a component experiences an error significant enough to have a system-wide impact but not significant enough to prevent the component from properly reporting its condition. Another issue with NACKs is that they may decrease the overall security of a system by making receivers infer a functioning system in the absence of communications [2].

To understand how acknowledgement schemes assist in detecting system-level faults, some analysis of the nature of faults is required. The field of computability theory defines the halting problem [3] as the problem of deciding whether a given program will halt[1] given a specific set of inputs. Counterintuitively, the general formulation of the halting problem is undecidable [4]; there is no way to ensure that a given program will always complete. If there is no guarantee that a component will complete its task, heuristic methods must be used to infer the likelihood of a successful outcome. One such method is to simply let the program run for a finite amount of time, and if the program does not complete within that time, assume it will not ever complete.

The difficulty with applying a wait-and-see heuristic is calculating the appropriate amount of time to wait for a response. The field of statistics defines the optimal stopping problem [5] as the problem of deciding when to take an action to achieve some optimal outcome. Applied to the fault detection problem, optimal stopping would involve determining when to take the action of declaring an error rather than waiting for a component to return a data result. Approaches to solving optimal stopping problems require some probabilistic analysis of future events; the more future events can be quantified, the easier it becomes to provide a good answer to this problem. One technique to reduce ambiguity relating to future events is to engineer timing thresholds into the system. Components will be required to produce results (or intermediate results) at specified time intervals. Because these intervals are regular, the only ambiguities in the system relate to the communication of results and any boundary conditions related to when results are sampled.

The technique of waiting a fixed time for information, and inferring failure in the absence of information, is formalized in the concept of a watchdog timer. Once set, a watchdog timer is decremented at regular intervals until it reaches zero, at which point it expires. When the timer expires, one or more of the components under the purview of the watchdog are considered to have failed and corrective actions need to be taken. Components can prevent the watchdog from expiring by resetting[2] the watchdog timer back to its original value, thus postponing its expiration. Because this reset only delays expiration, it must be done at regular intervals.

Watchdog timers can be used for a single component or across multiple components and can be implemented in software, in hardware, or both. A common approach is to have a software watchdog whose expiration action is to restart one or more software processes and a hardware component whose expiration resets the hardware itself. The software watchdog communicates with the hardware watchdog, ensuring that if the software watchdog itself stops functioning, or encounters an error it cannot fix, there is a hardware solution to apply a correction. When applied to the fault detection problem, watchdog timers provide a reliable way to

---

1. A program might not halt if it goes into an infinite loop.
2. The act of resetting the watchdog timer is informally known as petting the watchdog.

specify the behavior of a functioning system, and therefore are an unambiguous mechanism for detecting error conditions.

While watchdog timers typically use ACK schemes, it is possible to implement a watchdog timer that uses a NACK scheme (see Figure 13.1). In that case, the watchdog will constantly reset its own value and only expire on the receipt of one or more NACKs from a system.

The NWD pattern provides a mechanism for implementing a watchdog concept across portions of a DTN. In this pattern, nodes in a network serve as system components that check in with a network watchdog. System components use an ACK-based watchdog and a network or networking region uses a NACK-based watchdog. The network watchdog can apply corrective actions through network-wide configuration.

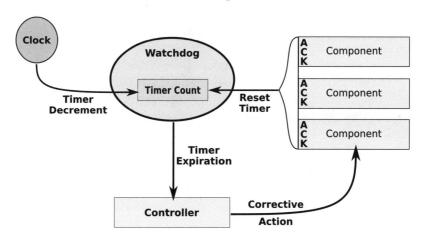

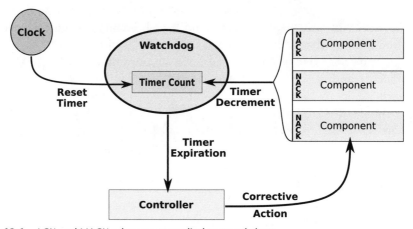

Figure 13.1 ACK and NACK schemes as applied to watchdogs.

13.2 The Problem Being Solved

The computational insight behind the design of a watchdog system is that there is no guarantee that a given component will ever complete its computation. Certain inputs may cause software to enter infinite loops. Computers may experience memory errors or reset as a result of transient events caused by operating in a harsh environment, such as the high radiation environments encountered by spacecraft. Aging systems experience hardware faults that cause unexpected (and perhaps recurring) reboots. Given these ambiguities about the reliability and resiliency of individual components in embedded systems over time, ACK-based watchdogs are used to detect and ultimately correct errors.

While the watchdog concept seems relatively simple, there are multiple design decisions that must be made correctly to ensure useful behavior. Most of these decisions relate to the quantification of various delays in the system. Improper consideration of delays may lead to false positives and false negatives, both of which lead to ineffective and potentially harmful corrective actions (see Figure 13.2).

A false negative indicates that a failure occurred in a system component but was not detected by the watchdog timer. While it may appear that watchdog timers are immune to this type of error as a result of using positive acknowledgement, there are conditions that lead to false negatives. First, watchdog timers can be fooled if a past acknowledgement is mistaken for a current acknowledgement. This condition is most likely encountered when the network used to communicate acknowledgements mistakenly sends an ACK more than once. Embedding unique information (such as timestamps or sequence numbers) into an ACK may prevent this problem, but at the cost of requiring watchdogs to keep history, increase processing time, and receive a larger data volume of larger ACK message sizes. Second, if a watchdog timer detects an error and immediately begins to implement a corrective action then a second component in error may not be reported because the watchdog has already initiated a recovery sequence from the first error. Waiting for one or more reporting cycles to ensure that all errors have been detected delays recovery responses in ways which can cause further risk to the system.

A false positive occurs when a watchdog timer infers that a component has encountered an error when it has not (see Figure 13.3). This condition only occurs when the watchdog timer expires—meaning that an ACK from a component failed to be received. However, failure to receive an ACK does not necessarily mean that the ACK was never sent as there could have been an issue with the network used to communicate the ACK. Initiating a set of system-wide corrective actions may result in significant impact to the system. When corrective actions are applied against a system that is not actually experiencing an error, the health of the system may be in jeopardy. Preventing false positives requires extending the expiration time of the timer to allow for a longer period of time to accumulate information. A timer can be increased as a function of predicted signal propagation delay and then further extended to account for multiple round-trips to allow for cases where network recovery requires retransmitting an ACK.

Watchdog timers discover system-impacting faults in a timely fashion by making specific assumptions on how components self-assess and communicate. Communications mechanisms between components and the timer must be reliable and deterministic, lest errors in the communications medium be mistaken for errors in

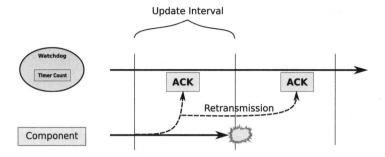

False Negative - Extra ACKs

If an acknowledgement is allowed to persist in the network it may outlast the life of the component which generated it. For example, if an acknowledgement is sent twice because of an issue with retrandsmission timeout, it may last longer than the component itself. This will cause a watchdog timer to fail to expire when it should.

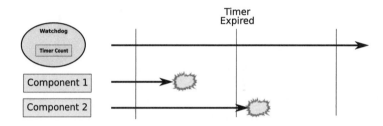

False Negative - Premature Expiration

If a watchdog timer expires because of one component failure, but other components have failed simultaneously, then the failure of those subsequent components might not be detected or reported by the watchdog.

Figure 13.2 Not considering delays can lead to false negatives in a watchdog.

the component using the medium. Using watchdog timers with spatially distributed components over common networks requires end-to-end analysis of networking performance and worst-case delays inherent in the system. If the characteristics of the network change significantly post-analysis, the timer system may become unstable and suffer from false positives and false negatives.

The benefits of a fault-detection mechanism would be useful for partitions of a DTN that would otherwise be unable to detect faults in the more traditional architecture of pushing performance and health data back to a centralized network operations center. A mechanism is needed by which faults could be detected, and corrective actions initiated, in partitions of a DTN.

13.3 Pattern Overview

The NWP provides a structure and set of behaviors for implementing a watchdog system distributed across portions of a DTN. In doing so, the pattern provides a

13.3 Pattern Overview

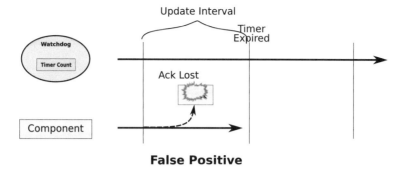

False Positive

If a positive acknowledgment is sent by a component, but not received by the watchdog, then the timer may expire and initiate a corrective action. Acknowledgments can be missed if they are lost in a network, corrupted, or sent by a component at the end of an update interval.

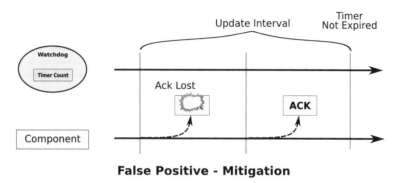

False Positive - Mitigation

Extending the update interval across multiple intervals gives more time to collect acknowledgments from components and avoid false positives. Doubling the update interval allows for the possibility of retransmissions and the case where an acknowledgment is sent right before the end of one interval.

Figure 13.3 Delays and disruptions can lead to false positives for watchdogs.

distributed fault-detection mechanism that can be used to help apply corrective actions beyond the scope of a local node. Because DTNs cannot guarantee timely, reliable data communications the implementation of this watchdog pattern must include both central and node-local responsibilities. The pattern is comprised of four logical responsibilities: a man1aged locality watchdog (MLW), local watched components (LWC), local node watchdog (LNW), and a managed locality controller (MLC).

Operating a watchdog timer across a DTN makes it impossible to implement the kind of determinism in the communications layer necessary for lack of data to imply component failure. Additionally, once a fault is detected, applying a corrective action over a DTN adds further ambiguity regarding when those actions will be received and applied. To address these problems, the concept of a network watchdog must define both a managed locality and abandon positive acknowledgments amongst nodes.

A managed locality is defined as a subset of nodes in a DTN that have more determinism in their communications. There are several examples of how such localities can be created. They may be created when two networks are federated into a single network, where each original network serves as its own locality. In cases where sensors communicate their information to a single cluster head, the sensor cluster may represent a locality. Even cases where multiple planetary rovers communicate to planetary orbiters can create localities—while communications may not be instantaneous, they are plannable and therefore deterministic.

Even when communications are deterministic, positive acknowledgments can produce too large a traffic volume to serve a networked watchdog function. A balance between positive and negative acknowledgments is to implement a two-tiered hierarchy for acknowledgments: positive acknowledgments for watchdogs running on local nodes and negative acknowledgments for the network watchdog (Figure 13.4). This approach allows individual nodes to perform some local fault detection and only communicate verified problems to the network watchdog, thus reducing the overall amount of traffic over the network.

13.3.1 Role Definitions

The MLC exists on the node in the locality that has the capability to apply corrective actions. This controller will have lower-latency access to nodes in the locality and serve as the central point of knowledge for how system-level corrective actions should be applied. It is the responsibility of the MLC to apply the correct actions to the correct nodes in the correct order at the correct time (Figure 13.5).

LWCs reside on those nodes participating in the managed locality. Component in this context may refer to a software executable, software process, or some device attached to the local node via a local, reliable network. Not every software process or attached device is an LWC. The decision to watch or not watch a component is based on an analysis of whether the failure of the component will have an impact beyond the node on which it resides. Those components whose processing enables or impacts the processing of the managed locality—or whose correction requires changes to other nodes in the managed locality—should be tagged as LWCs. When a component is identified as being an LWC, it must periodically, positively acknowledge its proper function to its LNW.

Local node watchdogs (LNWds) reside on nodes in the network participating in the managed locality. Similar to other types of watchdogs, LNWd timers are updated at regular intervals either to monotonically decrement to zero or to be reset to an initial value. To prevent false negatives, the update interval for the LNWd is calculated to be long enough for all LWCs to be able to provide a positive acknowledgement under nominal operating circumstances. At the end of the interval, if any LWV failed to provide a positive acknowledgement, the timer can decrement. If all LWCs provided their acknowledgements, the timer is reset to its initial value. If the LNWd time expires then a local corrective action may be taken, and a list of LWCs that failed to check in are sent as negative acknowledgements to the primary and backup watchdogs.

The MLW exists either on the same node as the MLC or on a node with reliable, low-latency access to node containing the MLC. As a type of watchdog, the MLW accepts acknowledgements from watched components, and if an error

13.3 Pattern Overview 227

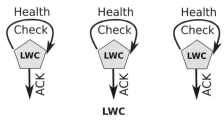

LWC

Components in the system that positively acknowledge their proper function to the LWC. The health of the LWCs are used to infer the overall health of the activities in the node.

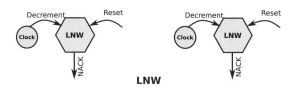

LNW

A watchdog time running on a node which requires AKS from LWC coresident on the node. If an LWC fails to check in, a NACK is sent to the MLW.

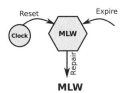

MLW

A negative watchdog which resets itself and expires if a NACK is received from a LNW. When expired, the MLW requests the MLC to implement a corrective action in the network.

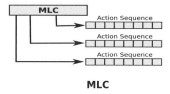

MLC

When kicked off by the MLW, determine an appropriate corrective action and implement it in the proper time and proper order.

Figure 13.4 Network watchdog system components.

condition is necessary, it communicates with the MLC to initiate a corrective action. The MLW operates differently from other watchdogs because it must collect information from nodes spatially distributed across the managed locality. Rather than requiring a regular stream of positive acknowledgments, the MLW timer uses a negative acknowledgement from any of the local watchdogs operating on any of its nodes. The MLW resets itself every period unless a NACK is received over the network, in which case it immediately expires at the end of the update interval, sending a list of received NACKs to the MLC.

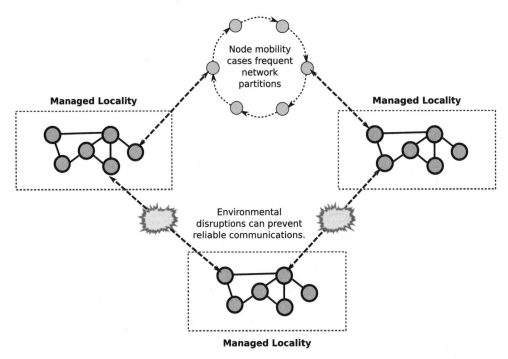

Figure 13.5 Managed localities can exist in challenged networking environments.

13.3.2 Control and Data Flows

The data flows through this pattern represent a one-way data flow from components to watchdogs to controllers (see Figure 13.6). The function of the controller presumably provides data back to watchdogs and controllers as part of applying corrective actions, but that is not part of the fault detection pattern.

The most challenging part of implementing an MLC is ensuring the ordering and timing of these actions. There are two strategies that can be applied in these situations to help in the design of a specific MLC: Timed-tagged commanding and acknowledged serialization.

Time-tagged commanding refers to the practice of annotating a command with the time at which the command should be executed. These times may be relative to the current time (e.g., run the command 10 seconds from now) or may represent an absolute time (e.g., run the command next Friday exactly at 3 p.m. UTC). The benefit of annotating commands with relative times is that the time specification uses fewer bytes in the command message resulting in smaller message sizes. The limitation of relative timing is that there is no way to correct for signal propagation delays associated with the transport of the command message. If a command is to be run 10 seconds after receipt and it takes between 0–10 seconds for the command to eventually be received over the network, then there is no way to determine exactly when the command would be executed on the platform. Relative times are almost exclusively used in cases where the network communicating the commands is deterministic. The benefit of absolute times is that there is no ambiguity on when a command should be executed (assuming that clocks are correctly synchronized) whereas their single limitation is that an absolute time makes overall message sizes larger.

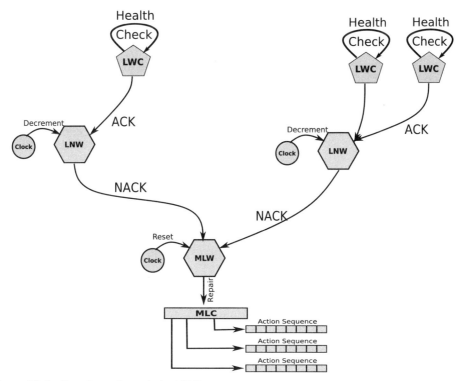

Figure 13.6 Data flows through the NWP.

Acknowledged serialization (see Figure 13.7) refers to the practice of not sending a command until some prior command has been acknowledged as being processed by some remote node. Processed can mean either received, queued for running at a future time, dispatched for execution, or successfully executed, depending on the overall design of the MLC and the nature of the corrective actions. Once a prior-sent command sequence is acknowledged, then a subsequent command sequence can be sent.

For example, an MLC may choose to coordinate a corrective action across multiple nodes in a network by sending time-tagged command sequences to each node in the network. The times in the command sequence would be tagged with absolute times initiated far enough in the future to increase the likelihood that all commands are received over the local area network prior to coordinated execution. Alternatively, the MLC could send individual command sequences to each node, wait for the node to complete it actions, before sending a sequence to some other node.

13.4 Service Types

The mapping of pattern components to service layers is driven by the most logical layers of the service stack to instantiate data repositories (see Table 13.1).

LWCs can exist in any service type, as any software application running on a node can be designed to provide a positive acknowledgement of health to some local watchdog process. Having LWCs at the lower service layers can provide helpful

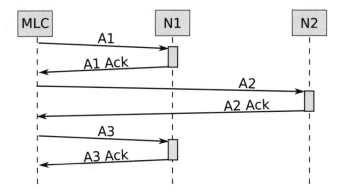

Serialized Commanding

The MLC sends commands in the order they are to be run by nodes. A command must be acknowledged before the next command can be sent. This gives maximum control over commanding, but relies the most on reliable, timely network communication.

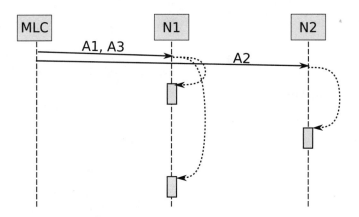

Time-Tagged Commanding

Commands are tagged with the time they are to run. The MLC sends batches of commands to each node and the node runs those commands in accordance with the time schedule. This reduces the overall traffic in the network and allows all commands to a node to be received at once. It relies on all nodes getting the time-tagged sequences.

Figure 13.7 Time-tagged commanding vs acknowledged serialization.

Table 13.1 Network Watchdog Services

Service	LWC	LNWd	MLW	MLC
I. Transport	X	X		
II. Networking	X	X	X	X
III. Federating	X	X		
IV. Overlay	X	X		
V. User Endpoint	X	X	X	X

insights into the ability of the node to participate in the managed locality, particularly in determining whether network throughput matches expectations. Because LNWds communicate directly with LWCs, they may also exist at multiple service layers.

The MLW and MLC are most effectively instantiated at the network and user endpoint layers. Implementing these functions at the network layer provides the ability of the watchdog and controller to apply network-wide corrective actions the operate beneath an overlay network. In cases where fault detection and corrective action need to operate using the abstractions provided by overlays and federations —or when dealing directly with application data—these roles should be applied directly at the user endpoint layer. In these circumstances, the MLW and MLC would exist as user applications operating on the overlay or federation, collecting any data that could be addressed to them by the respective LNWds.

13.5 When and How to Integrate

The NWD pattern can provide a useful architectural approach to fault detection in certain types of challenged networking environments. This section describes when this pattern should and should not be used and how to make design decisions on the mapping of pattern roles to software applications.

13.5.1 When to Use This Pattern

The NWD concept of a two-tiered, positive/negative acknowledgement watchdog is not common practice over terrestrial networks. The value of this approach for DTNs is that it balances the fault-detection abilities of positive acknowledgements with the lower bandwidth utilization of negative acknowledgements. Not every network architecture needs this type of balance, and the NWD should be considered when the following conditions are met.

There is a manageable locality in the network. Regardless of the acknowledgement mechanism, a watchdog timer must be able to detect faults such that corrective actions can be applied within some performance constraint. If the network providing communications to and from the watchdog is non-deterministic there is no guarantee that faults will be detected, or corrective actions initiated, in time. DTNs, by their nature, cannot guarantee communications across arbitrary nodes in the network. However, many practical deployments of these networks contain subnetworks, constituent networks, or otherwise network partitions that support more determinism then the rest of the network. Within this locality, distributed watchdogs can be implemented.

Network faults can be expressed as one or more node faults. The purpose of a network watchdog is to detect faults across some portion of the network. The NWD use of a two-tiered watchdog uses local node fault detection to inform the detect of network fault. This strategy only works if network faults are defined as one or more node faults. In cases where a network fault can exist but each node in a locality has no detectable fault, this pattern cannot be applied.

Coordinated corrective actions exist in response to error conditions. The benefit of detecting faults the happen across multiple nodes is that corrective actions can be distributed across multiple nodes in the network. This presumes the existing of multinode, coordinated, corrective action. If the responses to local node faults are local node corrections, then there is no need for a distributed corrective action.

Network communications within the locality are not high-rate or high-available. A manageable locality requires that communications be deterministic—not necessarily also high-rate and high-availability. In cases where a locality can support timely, guaranteed communications the NWP is not necessary; nodes can simply feed a single distributed watchdog using positive acknowledgments across the locality. If communications in the locality cannot support positive acknowledgement schemes either because the data rate is too low to support the data volume, or contacts are periodic, then negative acknowledgments must be used across the locality instead.

13.5.2 Recommended Design Decisions

The detailed design of components, and the way in which they should be integrating into a functioning network, are specific to the eventual operational environment. However, in generating the detailed designs of these components, multiple recommendations exist for how to make design decisions that result in a scalable, efficient, and operational implementation.

The LWCs and LNWs should be combined with local autonomy processes where possible. The interaction between LWCs and LNWs is similar to the interactions between the data collection and local control plane components of the TWP, and fault-detecting autonomy systems in general. Local nodes in any network need to perform some fault local fault detection and corrective action and these mechanisms should be used, where practical, to implement these portions of the NWP. In this case, the corrective action of a detected, unrecoverable node fault would be the sending of negative acknowledgement messages to the MLW.

Local Node Watchdogs should attempt to correct local errors before reporting. Not every local node fault requires a coordinated network response. LNWs should determine whether a local corrective action could be applied to address a fault in ways that overall network function would not be perturbed. For example, a local corrective action of resetting some software on the node may be preferred as an initial correction prior to sending negative acknowledgments to the MLW.

Local Node Watchdogs should annotate NACKs. The inclusion of additional, annotative data in NACKs from the LNW to the MLW help determine the best corrective action that can be applied across the network. At a minimum, the time the fault was detected and the application(s) that failed should be included in the NACK. Time information is useful to determine the likely impact of having the fault persist for the period of time between detection and application of a corrective action. Application information is important when the failed application provides configuration and status information for the network as part of its control plane. In such a case, not only are message flows through the failing node at risk, but also the continued operational awareness of the nodes fed control-plane information from the failing node.

The MLW should use a multi-update-interval window to collect NACKs. Systems engineering must be performed when determining how to configure the update intervals associated with all watchdog timers. One consideration in his timing analysis is the tolerance to errors and boundary conditions in the generation of NACKs to the MLW. If the update interval is defined as the time it should take for a message to be communicated from the LNW to the MLW, then any need for retransmission would prevent the NACK from being received within a single update interval. Similarly, a fault may be detected in one update interval and not reporting until the subsequent reporting interval. When the MLW determines whether multiple NACKs from the same node or from different nodes represent a correlation it should look across a window of multiple update intervals.

Corrective actions should be bundled to prevent delivery issues. The primary responsibility of the MLC is to apply corrective actions in the proper order and the proper time to those nodes that require corrective action. Attempting to supervise the running of commands across multiple nodes is challenging enough in high-availability, high-rate networks. Because there is no guarantee for such communications between the MLC and local nodes, strategies to reduce ambiguity must be adopted. One such strategy is to send all commands destined for a single node as a single message. Each command can be time-tagged to provide synchronization on the local node as well as across multiple nodes in the network.

13.6 What Can Go Wrong

There are very few guarantees in highly challenged environments, making the deployment of this pattern susceptible to a variety of potential issues.

Local status changes faster than the MLC can react. It is possible that the fault detected by an LNW is the symptom of a larger problem on the node. For example, memory corruption or processor failure on a node could certainly require local fault autonomy to take a local corrective action after a NACK was sent by the LNW. If a node reconfigures itself, reboots, or switched processing to backup components, the network-wide corrective action undertaken by the MLC may not produce the expected result. Wherever possible, designs should prioritize local node corrective actions over allowing the LNW to expire. Further, network corrective actions from the MLC should, whenever possible, request nodes to get to a known state by their own local means to help minimize the impact of a local node state change.

Communications with the MLW are affected by the fault. The danger of any NACK scheme is that the fault which causes the NACK to be generated may also prevent the NACK from being sent. If a node experiences a catastrophic failure it will likely be unable to send a NACK, and therefore unable to instigate a network-wide correction. In this case, if the node's inability to operate truly represents a network-wide issue, it is likely that other nodes in the network will, eventually, trigger their own errors as part of an error cascade resulting in a delayed set of NACKs coming to the MLW. The failure in this case is not in the ability to detect a fault, but a failure to detect a fault early enough to prevent a cascading effect in the network.

The MLC is on the faulted node. When the node that fails also includes the MLC, there are two important ways in which recovery is compromised. First, if the MLC is unable to run, or communicate over the network, it cannot send corrective actions into the network. Second, if the MLC is not functioning, NACKs sent to it could be lost such that even if the MLC is able to regain function later, the initial error reporting could be lost. These two conditions can be mitigated by supporting multiple MLCs in a locality, such as a primary and a backup, although this creates a synchronization issue. Alternatively, LNWs may periodically resend NACKs to avoid conditions where a NACK may be lost.

13.7 Pros and Cons

This section summarizes the benefits and pitfalls of using this pattern (see Table 13.2).

13.8 Case Studies

This section provides one examples of this pattern applied to administrative records associated with the BP.

13.8.1 BP Administrative Records

BPAs can send administrative reports to specified nodes in the network in response to certain conditions local to the BPA. While any node can be the recipient of an administrative message, in practice these messages would be destined for a small set of administrative nodes used to collect and analyze the reports. One type of administrative report associated with the BP is a bundle status report, which is generated whenever there is a failure by the BPA to process a bundle for some reason (see Table 13.3).

The situations under which the BPA sends these reports is a matter of the per-node policy configuration of a node. While always generating these reports

Table 13.2 NWP Pros and Cons

Design Goals	Pros	Cons
Detect Local Node Failures	Local watchdog timers provide fine-grained error detection on the node.	Deferring corrective action to an MLW delays fixing local problems.
Communicate faults to a network-central watchdog	Using NACK strategies allows for expiring a network watchdog timer without increasing network utilization.	NACKs might not make it off the local node if the failure is catastrophic.
Determine network-wide corrective actions	A central MLC can collect NACKs from nodes in the network and use the information in them to apply corrective actions.	NACKs may trickle into the MLC across various update intervals making the construction of an appropriate corrective action difficult.
Apply corrective actions in proper order at proper time	As a single entity, the MLC can send time-tagged batch commands to nodes causing coordinated, timed recovery.	Messages from the MLC may be delayed beyond their time-tagged delays, or lost in the network, causing the network to further lost synchronization.

Table 13.3 BPv7 Status Report Reason Codes

Reason Code	Meaning
0	No additional information.
1	Lifetime expired.
2	Forwarded over unidirectional link.
3	Transmission canceled.
4	Depleted storage.
5	Destination endpoint ID unintelligible.
6	No known route to destination from here.
7	No timely contact with next node on route.
8	Block unintelligible.
9	Hop limit exceeded.

produces a good amount of diagnostic information, it can also easily congest links. In cases where a node chooses to send a status report because of some failure of a process on that node to complete in a timely fashion, then the BPA sending the status report is acting as an LNW and the send status reports represent NACKs in the NWP.

For example, a BPA may be unable to store a bundle until its next forwarding opportunity. Lack of storage on a node is clearly an issue in a store-and-forward network and this condition is likely to affect other nodes in the network as well. If other nodes are using the full node their forwarding strategy will fail and cause the originating nodes to incur retransmission costs. Further, if the full node was filled by a message storm of some kind, it is possible that other nodes in the network may also have their local storage filled or will have their storage filled if several bundles on the full node are waiting a forwarding opportunity to a downstream node.

The BPA bundle status reporting mechanism follows the NWP as follows (Figure 13.8). The BPA in this case acts as both an LWC and LNW. As an LWC, the BPA provides an indication of fault on the node; in the aforementioned example, the fault being the inability to store a bundle. As an LNW, the BPA makes the determination that this local failure may require a network-wide response. Presumably, if the local node's storage has been exceeded, and this condition persists, it could affect multiple message flows in the network. In this case, the administrative message serves as the NACK in the system. The administrative node acts as the MLW, and potentially, the MLC. As the MLW, the administrative node receives the NACK message to detect a potential failure condition. As the MLC, the administrative node may also work to apply a corrective action, such as adjusting contacts or propagating updating congestion information through the network.

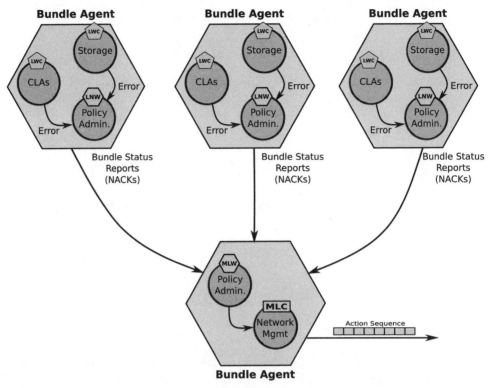

Figure 13.8 Processing bundle status reports follows the NWP.

13.9 Summary

The NWP provides a mechanism for detecting faults across nodes in a managed locality. The managed locality represents an area of deterministic (but not necessarily high-rate or high-availability) communications that is otherwise disconnected from traditional network operations by delayed or disrupted links. There is an advantage to detecting network faults quickly so as to increase the amount of time in which a corrective action can be applied. The faster corrective actions can be applied, the less chance that a single node failure will cascade into more errors across the locality.

A popular way to detect faults early in a system is to implement a watchdog timer upon whose expiration a corrective action is initiated. Processes local to the watchdog timer must continually reset the watchdog timer, thereby asserting their proper function. This mechanism of constantly asserting health is called ACK and provides a very reliable way of detecting faults. Attempting to implement this over a network may require more throughput than is available and will not work correctly when communications are delayed or deterministically disrupted. An alternative approach is to communicate only assertions of failure as NACKs on the presumption that failure conditions are less common than success conditions.

The NWP pattern implements a local, ACK-based watchdog timer whose corrective action upon expiration is to send a NACK to a locality-wide watchdog timer. This combines the high-rate local fault detection of an ACK scheme but saves the bandwidth over the network by using a NACK scheme. The network-wide watchdog resets itself continually unless it receives a NACK, and which point it works with a locality-wide controller to apply corrective action. The implementation of this pattern is very application-specific, but one trivial example is found in the BPA administrative reporting mechanism.

13.10 Problems

13.1 Provide three examples of an embedded system and explain why they should be considered embedded.

13.2 Provide a description of a fault that can result in a system-wide failure if not detected in a timely fashion. Explain how the fault would be detected, what kind of corrective action would need to be applied, and how that action could be applied in the correct order at the correct time.

13.3 Provide an example of a fault that might not be detected by an ACK scheme. Provide an example of a fault that might not be detected by a NACK scheme.

13.4 The halting problem states that there is no guarantee that a component (program) will halt (complete). Consider a simple program that prompts a user for two numbers, adds those numbers, writes the sum to a file on disk, and then exits. Provide three things that could prevent this simple program from completing.

13.5 Describe the use of a watchdog task to determine whether a heater has silently failed to operate on a spacecraft. What components would need to check in with the watchdog, how would the expiration time be calculated, and what corrective action should be taken? You may hypothesize any system design or set of components you feel is helpful.

13.6 Provide a realistic example of a false negative caused by extra ACKs in a system. Provide a realistic example of a false negative caused by premature expiration of a timer.

13.7 Provide a realistic example of a false positive and how extending the update interval for the timer would prevent the false positive.

13.8 Consider any of the challenge network examples used in the prior chapters and provide an example of a managed locality that could exist in that network. What makes the managed locality exist, and how could it be defined by network architects?

13.9 Describe a corrective action sequence that requires coordinating multiple actions across three nodes in a network such that actions on the nodes must be interleaved in time. Explain how a time-tagged approach can be used, and how an MLC can calculate the times to use.

13.10 Describe what happens in a time-tagged system if each node has a different send of time, or an unsynchronized clock. What time representations could be 0used in the time tag to address changes in the clock time at different nodes?

13.11 Provide an example of a distributed system that would benefit from the use of the NWP. In doing so, describe how the distributed system meets all of the criteria for using this pattern.

References

[1] Macker, J. P., and R. B. Adamson, "A TCP-Friendly, Rate-Based Mechanism for NACK-Oriented Reliable Multicast Congestion Control," *GLOBECOM'01, IEEE Global Telecommunications Conference (Cat. No. 01CH37270)*, Vol. 3, IEEE, 2001.

[2] Compagno, A., et. al., "To NACK or Not to NACK? Negative Acknowledgments in Information-Centric Networking," *2015 24th International Conference on Computer Communication and Networks (ICCCN)*, IEEE, 2015.

[3] Turing, A. M., "On Computable Numbers, with an Application to the Entscheidungsproblem," *Proceedings of the London Mathematical Society*, Vol. 2, No. 1, 1937, pp. 230–265.

[4] Rybalov, A., "On the Strongly Generic Undecidability of the Halting Problem," *Theoretical Computer Science*, Vol. 377, No. 1–3, 2007, pp. 268–270.

[5] DeGroot, M. H., *Optimal Statistical Decisions*, Vol. 82, John Wiley & Sons, 2005.

CHAPTER 14

The Data Forge Pattern: Leveraging In-Network Storage

This chapter presents a mechanism by which applications in a store-and-forward network can leverage in-network storage for tasks such as data fusion and increased network reliability. Using this mechanism, waypoint nodes can increase the likelihood that destinations receive only useful information. Removing duplicate or otherwise irrelevant information reduces the overall traffic volume in the network.

The data forge pattern (DFP) describes how nodes that implement a store-and-forward message cache can reduce duplication and eliminate expired information. The pattern is named after the concept of forging as a manufacturing process whereby new alloys can be created by heating and combining two or more other metals. A data forge being a process by which one or more data sets are combined to create a new data set.

14.1 Pattern Context

Nodes participating in a store-and-forward network use persistent storage to hold incoming messages while waiting for an appropriate egress to the message's next hop. When gaps in access to these egresses are infrequent, this storage acts as a rate buffer providing temporary relief to network backpressure. When gaps in access to egresses become frequent and/or persist for longer periods of time, messages accumulate in persistent storage and may lose their timeliness and relevance. The only way to truly congest a DTN is to exhaust its persistent storage, and several approaches have been proposed to preserve storage options in these networks [1–3].

The DFP provides a mechanism by which applications can utilize storage at waypoints in a sensing network to verify and/or consolidate data. The insight behind this pattern is that, by verifying application data prior to its delivery, the efficiency of the network can be increased because invalid data can be removed early. To discuss where this approach makes sense it is important to understand the points of view associated with application data in a network. Application data can be viewed as a stand-alone information element, an object consuming network resources, or a part of a larger user function. Together these three views form a tripartite model (see Figure 14.1) of information distribution: the information source view, the transport network view, and the user function view.

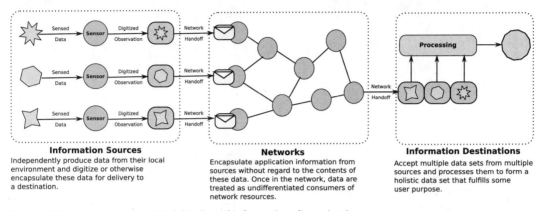

Figure 14.1 A tripartite model of distributed information dissemination.

Information sources, such as those in sensing networks, create stand-alone data that represents their knowledge at a point in time. The source may not understand the overall system in which it operates or the utility of its provided information. The information may be generated at regular time intervals or be event driven. Regardless of how and when the information is generated, the source is solely concerned with ensuring that the information is correctly captured and passed to the network for communication to some destination.

Network resources are shared by multiple sources in the network and treat source information as payloads annotated with network-specific data. These payloads, once placed in a networking wrapper, are not examined by the network again[1] because the networking annotations specify all the information necessary to secure, route, and otherwise deliver the payload to its destination. The network is completely abstracted from the purpose of the source information, why it was sampled, and what features the information enables. Instead, the network message is seen as a consumer of networking resources that must, by necessity, eventually be removed from the node because the message has been transmitted or deleted. (Figure 14.2) Information destinations accumulate information from one or more sources for the purpose of fulfilling some user function. The set of all such destinations comprise the purpose for which the network exists.

To see how this tripartite model is implemented in a simple information dissemination network, consider a system which calculates the ambient temperature of a building for the purpose of automating a climate control system. In such a network, individual sensor nodes sample temperatures frequently and digitize those measurements for transmission over the network. These nodes have no concept of the overall building temperature or how their temperature measurements will be used for climate control. Their point of view is to generate and send data assuming other nodes will make proper use of it.

The network transporting these measurements considers them as opaque payloads annotated with network information such as destination, priority, and time-to-live (which is often implemented as a hop count and not an actual time in the network). The network has no mechanism to handle the fact that a measurement

---

1. This is particularly the case in high-volume networks where deep packet inspection provides a computational bottleneck [4].

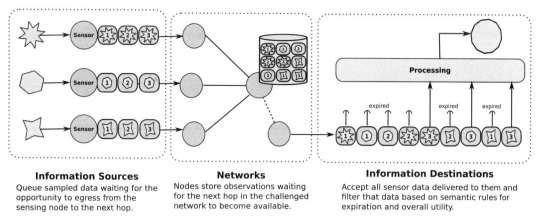

Information Sources	**Networks**	**Information Destinations**
Queue sampled data waiting for the opportunity to egress from the sensing node to the next hop.	Nodes store observations waiting for the next hop in the challenged network to become available.	Accept all sensor data delivered to them and filter that data based on semantic rules for expiration and overall utility.

Figure 14.2 The mapping of information to messaging in challenged network environments.

might only be useful within a certain timeframe[2]. Because the network in this case is well-resourced, the not-useful data will be delivered anyway because the perceived consequence of doing so is negligible.

The climate control destination adopts a holistic view of the system by fusing individual temperature measurements into an overall picture of the climate of the building. From that overall picture, the destination determines courses of action that ultimately fulfill the purpose of the system: maintain a set ambient temperature. Because sensors report information more frequently than necessary (a technique known as oversampling), the loss of any individual measurement (or the receipt of an expired measurement) does not endanger the ability of the destination to create a credible temperature model. For this reason, sensor networks often use the UDP for fast but unreliable transmission to avoid the overhead of sessions and retransmissions; because there are so many measurements being transmitted, error handling can be implemented simply as a matter of statistics.

Even this simple model of information dissemination becomes more complex in a DTN. In networks where messages can frequently be delayed or lost, the strategy of oversampling information reduces the goodput of the network. Sending multiple samples across a low-rate or transient network increases link congestion without a commensurate increase to the overall knowledge at the information destination. Similarly, accumulating messages at store-and-forward nodes decreases available storage and applies network resources for messages that may no longer have utility.

The DFP provides a mechanism by which intermediate nodes in a network may evaluate application data existing in a store-and-forward cache to assess their semantic value prior to allowing them to be forwarded. In this pattern, application agents in the network evaluate information even when they are not the information destination. By pushing the semantic evaluation of data upstream the overall data volume that must be communicated to the destination is reduces and needless transmissions and storage eliminated (Figure 14.3).

---

2. While receiving a temperature measurement two hours late may contribute to some long-term metric analysis, the measurement is not useful for real-time climate control.

The Data Forge Pattern: Leveraging In-Network Storage

Expiring Older Data
Nodes can reduce traffic in the network by inspecting the times associated with data elements. If times are determined to be too far in the past, the information can be removed from the network.

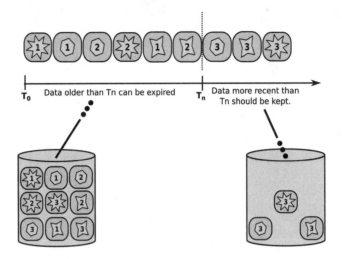

Fusing Duplicate Data
Nodes can reduce overall traffic in the network without losing any data by fusing data. In the case, if a value is sampled at multiple times without change, then a single message representing the time span can replace the individual, shorter measurements..

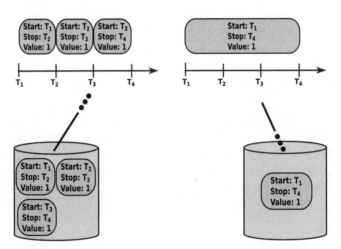

Figure 14.3 Data pruning and fusing in a DTN.

14.2 The Problem Being Solved

Information destinations exist to fulfill implementation-agnostic user requirements (e.g., calculate an average temperate every hour). The strategies used to implement these requirements depend upon the capabilities of information sources and the transport network. For example, the strategy of oversampling to ensure enough information reaches a destination assumes that resources exist in the network to communicate that data volume. DTNs do not always have the resources to employ this strategy. Put another way, oversampling does not increase the odds of delivery if no egress link is available from a node. The amount of stored data from an oversampling source can overwhelm the resources of low-rate links and storage on nodes.

A mechanism is needed by which nodes other than the destination can be used to filter message traffic. By pushing this semantic evaluation of application data upstream in the network, useless data can be rejected earlier, and existing data may be fused in ways that reduce the overall message volume.

Continuing with the example of temperature sensing, consider a source which produces temperature measurements every 10 minutes through a DTN to a central node which fuses that information to provide a common visualization of temperature variations over a region for the past hour. As these measurements traverse the DTN, they may be stored at individual nodes for long periods of time. In such a situation, a waypoint node may exist with multiple temperature measurements from the same information source and these measurements may be evaluated as being in one of three states: unique, expired, or duplicative.

If a measurement is unique amongst all other measurements on the node, then the waypoint would want to continue passing this information on to the information destination. If the measurement was generated more than an hour ago, it will not be a useful input to the information destination, whose requirement is to calculate an average over the past hour. In this case, the measurement might be a candidate for deletion so that it does not expend additional networking resources only to be discarded at the destination. Finally, if multiple measurements all show the same temperature, they may be combinable into a single, measurement spanning a longer timeframe, thereby keeping the same application information but otherwise reducing the overall number of messages (and thus, overhead) in the network.

By semantically validating information at waypoints in the network, and by fusing information where possible to reduce the overall data volume, the overall efficiency of the network can be increased. A mechanism is needed that can provide this type of application-specific semantic processing upstream in the network.

14.3 Pattern Overview

The DFP describes the logical components (see Figure 14.4) and their associated behaviors for implementing semantic processing of application data at rest on waypoint nodes in a network. By doing so, the pattern provides a mechanism to use computational resources on a waypoint node to save network resources by reducing data volume whenever possible. The pattern is comprised of five components: a local data cache (LDC), one or more information sources (IS), one or more information destinations (IDs), a semantic data validator (SDV), and a data fusion engine (DFE).

Data forging involves reviewing application-specific data stored on a local node for the purpose of ensuring the validity and compactness of the information prior to its future transmission. Ideally, this process would involve removing old data from the data cache and fusing existing information where possible. In cases

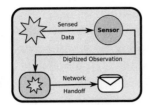

Information Sources
An information source is any networked node that generates data independent of other information sources in the network and sends those data to an information destination.

Local Data Cache
The local data cache stores in a store-and-forward network for future transmission. The data in these messages may be updated or removed as part of the DFP.

Semantic Data Validator
The semantic data validator reads application data from the local data cache and determines whether the data has expired or otherwise become invalid and able to be deleted.

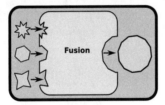

Data Fusion Engine
The data fusion engine examines valid data in the local data cache to see if any data can be combined into a new application data message to reduce the number of messages in the network.

Information Destination
The information destination is responsible for implementing some user function in the network and is the purpose of the application (and in some cases the network itself).

Figure 14.4 Data forge system components.

where messages are removed or resized, a local routing function may need to be rerun at the local node to recalculate a network path. If messages are removed, they no longer need to occupy capacity on future contacts. In cases where messages are made smaller, a sooner, smaller capacity contact may suffice for transmission. In cases where messages are made larger, the original contact allocated to the message may not have enough capacity to transmit the message.

14.3.1 Role Definitions

The LDC comprises the set of application data stored on a waypoint node and pending retransmission to some other node in the network. These caches may contain data as it was originally generated at an information source or some fused or summarized data created from multiple sets of application data. The data cache has two primary responsibilities for the DFP: proper storage of all application data in a way that is conducive for network transmission, and efficient identification and retrieval of application data for analysis. The most challenging part of implementing an LDC is efficiently searching application data sets stored on the node. Efficient searching of large data sets is often enabled by the creation of indices noting the location of data within a storage mechanism. In cases where there is a one-to-one mapping of messages to application data, the identification of where messages exist in the LDC may be enough.

ISs represent the nodes in a network that produce measurements that may either expire while in the network or that otherwise could be fused in the network prior to reaching their destination. ISs within the DFP are not presumed to have any semantic knowledge of how the data they produce will be used within the network or at the destination. In cases where an IS produces data which is then stored on its local node waiting for an egress, other logical components of the DFP may be used to expire or fuse that data. However, this does not imply that the IS has, or needs, additional knowledge of how information is used in the system.

The ID represents the consumer of the full set of data produced by the ISs in the network. From the perspective of the DFP, the ID functions less as a message destination and more as a source of requirements for how to determine the validity of information in the network.

An SDV determines the status of each piece of application data in the data cache, as to whether the information is unique, expired, or a duplicate. To perform this evaluation, the SDV requires semantic knowledge of the requirements imposed on the ID in the network and how to identify data held in the LDC. In cases where messages are considered expired, the SDV can request data removal from the LDC. In cases where messages are considered duplicates, they can be referred to a data fusion engine for processing.

A DFE operating on a local node evaluates potentially duplicate application data for the purposes of creating a new data fusion product that replaces one or more other data products in the LDC. Duplication can refer to exact copies of data resident on a waypoint node from duplicate transmissions or semantic duplication where multiple messages represent the same data (or the persistence of the same data over a period). In cases of syntactic duplication, the DFE can request that the LDC remove some number of duplicate nodes. In cases of semantic duplication, the engine can either create a new data set or update an existing data set to replace others in the cache. When the DFE completes its run, it may request that the LDC review changes data sets to see if new rout computations must be run.

14.3.2 Control and Data Flows

The data flows through the DFP are focused on the LDC as the storage of the data upon which the pattern operates. The complexity of the pattern is based on the mechanisms for efficiently locating application data in the LDC and determining expiration and duplication from the SDV.

As network messages are received at waypoint nodes, they are either immediately retransmitted, discarded by the node, or stored for future transmission. If a message is stored for future transmission, it is added to the LDC with some annotative information associated with when and how the message should be retransmitted when a future contact opportunity arrives. When information is added to the LDC, it is naturally indexed so that the message can be found at the appropriate time in the future.

To examine elements for semantic value, the SDV must be able to read information from the LDC and send notices to the DFE and LDC based on whether expired or duplicate information is found. The DFE receives notifications of duplicate data items from the SDV, implements fusion, and then provides new data back to the LDC for storage. In all cases, the LDC may choose to remove expired information or allow it to persist and to accept fused data or reject fused data, all based on policy.

The flow of control through the components of the DFP is as follows (see Figure 14.5):

1. One or more information sources generate data which are provided to the network and eventually stored in an LDC on a node. The LDC may stores information as network messages but must find a way to semantically process the information in its persistent storage and provide access to this data by the other DFP functions.
2. The SDV reviews statistics associated with the LDC to see if there are sufficient amounts of application data to warrant a validation activity. If so, appropriate data sets are reads into the SDV to assess their continued validity.
3. If the SDV discovers data which can be removed from the LDC, then those recommendations for removal are passed back to the LDC, which may or may not remove the data, as a matter of policy.

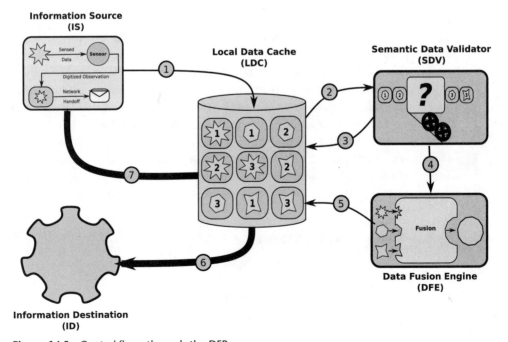

Figure 14.5 Control flows through the DFP.

4. If the SDV discovers data which can be fused, then the recommendation to fuse data can be passed to the DFE for processing.
5. The DFE may fuse data as it is able to, based on the information passed to it from the SDV. When completed, the DFE communicates the fused data, and data that can now be safely discard, back to the LDC.
6. The network will read information from the LDC to send along to the ID, as appropriate.
7. The LDC may send messages back to the ISs to ensure that removed data is no longer retransmitted, if necessary.

14.4 Service Types

The mapping of pattern components to service layers is driven by the most logical layers of the service stack to instantiate data repositories (see Table 14.1).

ISs and the IDs that they communicate can exist either at the network services tier or any services running over network services. The DFP is not applicable to transport services because these services communicate over a single-hop link, and therefore, there is no ability for intermediate storage.

Because the LDC stores network messages, it should be instantiated in places where network messages can be collected: at the network or overlay service layer. The SDV and DFE must be instantiated in a place that has ready access to the LDC, either at the same service layer or in the application layer, assuming the application layer has visibility into the cache.

14.5 When and How to Integrate

The DFP provides an architectural approach to exploiting common data storage to forge new data in the network and/or prevent expired or otherwise duplicate information from persisting in the network. The section describes when this pattern should and should not be used and how to make design decisions on the mapping of pattern roles to software applications.

14.5.1 When to Use This Pattern

The DFP premise of operations on a standard cache of information is not dissimilar from the use of CDNs, which cache information for easier retransmission. The

Table 14.1 Data Forge Service Types

Service	IS	LDC	SDV	DFE	ID
I. Transport					
II. Networking	X	X	X	X	X
III. Federating	X				X
IV. Overlay	X	X	X	X	X
V. User endpoint	X		X	X	X

concept of deep packet inspection on the content cache is not a common practice in networks because the consequence of sending expired or duplicate data are often not significant. However, this approach can be valuable in DTNs if it prevents wasted transmissions and needless storage. The DFP should be considered when the following conditions are met.

The data passing through the network is oversampled. The utility of fusion as a mechanism to reduce data duplication is strongly correlated to the likelihood of encountering duplicate data in the network. Duplicate data occurs when an information source periodically generates data even when the generated data is the same as some previously generated data. Examples of this kind of behavior include measurements from sensors and health telemetry from spacecraft, both of which are produced at regular intervals. In these cases, storage points in a network are much more likely to accumulate data representing semantic duplications in ways that can be more effectively consolidated as part of performing fusion.

Network nodes have enough processing power. The deep packet inspection required to determine whether there are semantic issues with the data in network messages requires significant computing resources. The LDC must be evaluated at regular intervals, and this evaluation cannot take so long that new transmission opportunities are lost because contacts become available before some pass through the LDC, SDV, and DFE cannot complete.

Data can expire in the system prior to the message containing the data. Protocols use a variety of methods to track how long a message may exist in a network, such as hop counts and time-to-live[3]. In certain cases, the data in a message may stop being useful before these message indicators require the message be removed from the network. For example, if an information destination only requires the most recent five data samples from a particular information source, and an LDC contains six such samples, the oldest sample can be removed even if the message holding that sample has not expired. When applications can define semantic data expiration criteria separate from the criteria used to remove messages from the network, this pattern should be considered.

Data congestion is a problem in the end-to-end network. The benefit of the DFP is that it can reduce data volumes in certain circumstances, leading to less contention for network resources. If the network is unlikely to experience congestion, then there is no need to undertake the complex process of identifying expired and duplicate data. The DFP should only be considered in cases where network nodes can accumulate large amounts of data relative to the egress opportunities of messages from the node. This can happen when there are low-rate contacts, infrequent contacts, or both.

14.5.2 Recommended Design Decisions

The detailed design of components, and the way in which they should be integrating into a functioning network, are specific to the eventual operational environment. However, in generating the detailed designs of these components, multiple

---

3. IP networks treat time-to-live (TTL) fields as a hop count, whereas DTNs use TTL fields as the actual timeframe in which the bundle is considered viable.

recommendations exist for how to make design decisions that result in a scalable, efficient, and operational implementation.

The LDC should maintain multiple indices to reduce data access times. As a central point of storage, the LDC is involved in all the data exchanges associated with this pattern. Special attention must be given to the implementation of this data cache to ensure efficient access of information. One way to reduce data access times is to create indexes of data locations as part of the initial storage of information in the LDC. In cases where there are multiple ways to look up information, there should be multiple indices maintained by the LDC.

The LDC should establish a locking mechanism to avoid data corruption. Accesses into the LDC must be protected such that a set of information being operated upon does not become corrupted by multiple writers. This limitation is true whether the LDC is implemented as a simple data buffer, an embedded database, or some other mechanism. The need to protect data in the LDC is particularly important when running the DFE which, by definition, fuses multiple messages into a single message. Until the DFE completes and the LDC applies its changes, none of the messages being fused should be read by any other process, particularly any process that would extract them from the LDC for transmission.

The LDC must consider the impact of removing messages from the system. The impact of message removal from the LDC, either because they are expired or because they are fused with other data, must be carefully considered. In cases where information sources oversample information and provide it to the network as best effort, the removal of messages has no consequence. The same cannot be said of cases where information sources require acknowledgements or otherwise maintain the possibility of data retransmissions. If the removal of a message incurs the possibility of a retransmission from the information source, the LDC must perform some action to prevent this, either by sending an acknowledgement back to the information source or utilizing some other transport protocol mechanism.

The SDV and DFE should operate solely on the information contained in messages. Where possible, semantic evaluation of messages should not require configuration of the SDV or DFE themselves. In a DTN, configuration updates cannot be guaranteed to be distributed across the network equally. As such, attempts to configure the network may result in the SDV and DFE on one node having a different configuration as those on another node, leading to inconsistent results. Defining semantic meaning in relative terms, such as last five samples are relevant enforces meaning without requiring configuration and should be preferred. Alternatively, absolute terms, such as any data older than Friday at 3 p.m. is to be expired, require regular updates and should be avoided.

The SDV should only run when the LDC achieves some critical mass of application information. In addition to maintaining indices, the LDC should keep statistics associated with stored application information. Statistics such as the total number of application data sets, average age of data, and the number of bytes of application data present can be used to determine whether the SDV should be run at all. For example, if there is only one piece of application data in the LDC, there is no likelihood of finding a duplicate set of data or expiring a significant volume of data.

14.6 What Can Go Wrong

There are very few guarantees in highly challenged networking environments, making the deployment of this pattern susceptible to a variety of potential issues.

Attempts to remove unnecessary data can lead to the retransmission of that data. If the LDC removes a message whose delivery is guaranteed by the information source, then the information source will retransmit that data if it thinks the data never reached the information destination. In cases where the LDC fails to send some upstream acknowledgement to the information source, it is possible that the message will be removed by the LDC of a waypoint, regenerated by the information source, removed again by the LDC of a waypoint, and so on until the overall time to live of the message is reached.

Locking the LDC can provide processing delays and data loss. The locking semantics of the LDC must be fine-grained or else the overall processing of the data cache will lead to LDC access delays and possibly dropped data. Because the LDC represents the persistent storage of the node, if an incoming message cannot be stored in the LDC it can only exist in a temporary buffering space; when that temporary buffering space is needed by a subsequent incoming message (or the incoming message is larger than the buffering space) the data will be dropped by the node. The LDC locks must be fine-grained to restrict access to only the messages being operated on by the SDV and DFE and to always allow new data to be written into the system (Figure 14.6).

LDC changes may cause cascading updates. The removal of messages from the LDC as part of expiration or fusion can free up so much anticipated future capacity in the network that all new messages in the LDC must be rerouted to see if they would benefit. This cascading update can cause large processing delays and negatively impact the accuracy of routing information propagated to node neighbors. An evaluation must be made as to whether an existing unmodified message required rerouting if the message already has a viable path to its destination to avoid this kind of evaluation cascade.

14.7 Pros and Cons

This section summarizes the benefits and pitfalls of using this pattern (see Table 14.2).

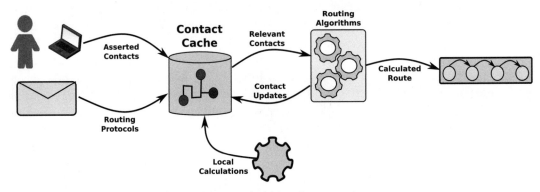

Figure 14.6 Local message caches are integrated with routing strategies.

Table 14.2 DFP Pros and Cons

Design Goals	Pros	Cons
Expire useless information.	Semantic evaluation of information detects expired data more accurately than time-to-live.	Processing semantic meaning for application data is time consuming.
		Semantic criteria may change or be configured differently across the network.
Remove duplicate data.	Data fusion provides a way to reduce the number of messages in the network without losing unique information.	Data fusion is timely and results in restricting access to portions of the LDC while the fusion is performed.
Minimize wasted storage at nodes.	Periodic removal of expired and duplicate data frees space in the LDC.	Indexes and statistics may generate more data than is saved when processing smaller data volumes.
Reduce wasted retransmissions	Removing the number of messages reduces the overall number of message transmissions in the network.	Failure to prevent information sources from retransmitting data can increase the overall number of retransmissions for a given message.

14.8 Case Studies

This section provides two examples of this pattern applied to store-and-forward networking and information-centric networking.

14.8.1 Store-and-Forward Routing Applications

To understand how an application may operate on an information cache stored on a local node we investigate how routing algorithms in DTNs use cached information and what operations they must perform on that cache.

Routing algorithms in store-and-forward networks must understand the topology of the network over time and match that information against the destinations of messages queued on the node. The topology of a network can be modeled as a series of contact opportunities over time [5] where each future contact opportunity is annotated with the start and stop time of the contact. One popular variant of routing in a DTN is time-aware variations of Dykstra's algorithm [6].

The local, persisted cache of contact information can be updated as a node's understanding of the network topology evolves. The mechanisms used to update the contact cache depends on the architecture and capabilities of the network itself. Networks with periodic changes to their topology may generate their own expected contact opportunities strictly as a function of time or by operator configuration [7]. Networks with ad hoc node mobility may implement opportunistic contact discovery mechanisms and build stochastic models of topology. Regardless of whether contact updates are generated by operator configuration, onboard deterministic or stochastic algorithms, or just-in-time discovery, they can all be aggregated into a single cache of contacts from which a routing algorithm can operate.

New contact opportunities can be made by adding nodes into the network, reconfiguring existing nodes in the network, and/or discovering existing network nodes. In certain cases, these new contacts can have start times before other contacts in the contact cache, and as such, might provide more timely egresses for any messages stored on the node. Alternatively, contacts might be removed from the cache if their associated nodes have been removed from the network, reconfigured, and/or other trusted information is received which makes a node unfit for use. In these cases, any messages waiting for that contact must be rescheduled to use some other contact in the system. Finally, a node determines that it may never have a route that can deliver a message prior to its expiration time, the message may be removed from persistent storage.

There are several logical functions that must be implemented by a node in support of a contact-cache routing system. The functions operate on both the contacts that define the topology over time, and the messages that use those routes. As previously described, the contact cache can be updated by adding, removing, or updating existing contacts over time. Separately, the routing system may need to deconflict potential misconfigurations in the contact cache by combining or splitting contacts that exist between the same nodes but with overlapping times. For example, if a new contact between two nodes is added that straddles an existing set of contacts, then those three contacts can be collapsed into a single contact. Similarly, if a new contact encompasses all or part of an existing set of contacts, then the contacts can be trimmed or removed to avoid duplication. The routing application must also operate on the message cache to determine whether a message should be expired (in which case it will not need to be routed and will not reserve capacity on a future contact), rerouted (based on new contacts), or sent over a currently operational contact.

14.8.2 Information-Centric Networking

An information-centric network (ICN) is any network where application data are communicated (and consumed) as a function of their content rather than as a function of specific (and predetermined) destination. The goals of ICNs are particularly promising in the area of Internet of Things (IoT) partly because IoT devices represent event-driven, machine-to-machine data exchanges [8–10]. This implies that the network messages used in these networks are operated on by the network as a function of their semantic meaning rather than the syntax in a specific networking protocol header. For example, a network message could be addressed to anyone near a specific geographic location or anyone who is interested in the latest temperature measurements. Because data in this paradigm do not have a predetermined destination, it is necessary that the data be held in the network waiting for potential information destinations to manifest.

ICN nodes, by definition, must keep data stored within the network because the user(s) of the data may not be known in advance. Storing this data in the network requires nodes to understand when the data is no longer useful or when it can be merged with other data to save space or increase utility. The software running on an ICN node must be able to process the semantic meaning of the data beyond what might be captured in a network transport protocol header, and these messages may come from multiple information sources.

The DFP is a useful mechanism for ICNs as it provides the architecture for ICN nodes to perform their required functions. The SDV on an ICN determines whether any application data may no longer be relevant in the network—as a function of time, number of forwards, observed traffic, or other criteria—and a DFE may be used to combine data as new information is received by a node. In fact, the concept of in-network storage and content-based addressed requires these functions to act upon the semantic meaning of the data.

14.9 Summary

Many networks provide a mechanism for dealing with errors associated with network transport. Checksums and other integrity evaluations are used to ensure that the network did not corrupt or otherwise alter a message between its source and destination. Time-to-live and hop counters ensure that a message does not get lost in a network for longer than its estimated useful lifetime. However, none of these mechanisms provides a way of evaluating the application data itself when making these decisions.

DTNs, by definition, struggle to communicate data volumes in a timely and reliable way. Because of this, information must be stored in the network by the network. If the network is unable to evaluate the utility of application data at rest in the network, then it is at risk of applying precious networking resources for data that will be discarded as soon as it reaches its destination.

The DFP provides a mechanism by which network nodes can choose to evaluate the application data inside of messages in a message cache, often referred to as deep packet inspection. As a result of this inspection, messages may be removed from the network or fused with other messages. In either case, the purpose of the DFP is to reduce the overall data volume communicated to a destination as early as possible. In this way, the limited networking resources of nodes in a DTN can be applied only to that data that has the highest likelihood of being useful to the information destination.

This approach is most impactful in distributed information dissemination networks where information sources oversample their data to try and account for likely losses on their way to information destinations. In these cases, the probability of a larger message cache containing expired and/or fusible data are increased making the computational overhead of the DFP justified by the likely reduction in overall data volume.

14.10 Problems

14.1　Provide a real-world example of the tripartite model of distributed information dissemination. In this example, describe the roles of each of the components and explain why they match the roles defined by this model.

14.2　List and describe two reasons why having the network layer not understand the semantic meaning of data is a benefit. List and describe two reasons why understanding the semantic meaning of a message would be a benefit to the network layer.

14.3 Describe an alternative to oversampling information in a distributed sensor network. In what network architectures will this alternative work and not work?

14.4 In addition to expiring old data and fusing duplicate data, provide a third strategy to reduce data duplication through a waypoint node.

14.5 Explain two negative consequences of not rerunning a routing algorithm on messages in the LDC if the SDV removes many expired messages from the LDC.

14.6 Explain one positive benefit of not running a routing algorithm on messages in the LDC if the SDV removes many expired messages from the LDC.

14.7 One challenge in implementing an LDC is efficiently searching for, and retrieving, data. List two other difficult problems associated with implementing an LDC.

14.8 Provide a design by which an LDC can communicate with one or more ISs to ensure that deleted data is not retransmitted by the IS.

14.9 Describe why the DFP should not be used in systems where ISs do not oversample their data.

14.10 Describe how the LDC can be corrupted when using the DFP if there is not a locking mechanism on it.

References

[1] Seligman, M., K. Fall, and P. Mundur, "Storage Routing for DTN Congestion Control," *Wireless Communications and Mobile Computing*, Vol. 7, No. 10, 2007, pp. 1183–1196.

[2] Seligman, M., K. Fall, and P. Mundur, "Alternative Custodians for Congestion Control in Delay Tolerant Networks," *Proceedings of the 2006 SIGCOMM Workshop on Challenged Networks*, ACM, 2006.

[3] Hua, D., et. al., "A DTN Congestion Mechanism Based on Distributed Storage," *2010 2nd IEEE International Conference on Information Management and Engineering*, IEEE, 2010.

[4] Cong, W., J. Morris, and W. Xiaojun, "High Performance Deep Packet Inspection on Multi-Core Platform," *2009 2nd IEEE International Conference on Broadband Network & Multimedia Technology*, IEEE, 2009.

[5] Araniti, G., et. al., "Contact Graph Routing in DTN Space Networks: Overview, Enhancements and Performance," *IEEE Communications Magazine*, Vol. 53, No. 3, 2015, pp. 38–46.

[6] Seguí, J., E. Jennings, and S. Burleigh, "Enhancing Contact Graph Routing for Delay Tolerant Space Networking," *2011 IEEE Global Telecommunications Conference-GLOBECOM 2011*, IEEE, 2011.

[7] Birrane III, E. J., "Building Routing Overlays in Disrupted Networks: Inferring Contacts in Challenged Sensor Internetworks," *International Journal of Ad Hoc and Ubiquitous Computing*, Vol. 11, Nos. 2/3, 2012, pp. 139–156.

[8] Quevedo, J., D. Corujo, and R. Aguiar, "A Case for ICN Usage in IoT Environments," *2014 IEEE Global Communications Conference*, IEEE, 2014.

[9] Baccelli, E., et. al. "Information Centric Networking in the IoT: Experiments with NDN in the Wild," *Proceedings of the 1st ACM Conference on Information-Centric Networking*, ACM, 2014.

[10] Zhang, Y., et. al., "Requirements and Challenges for IoT over ICN," IETF Internet-Draft 2015.

CHAPTER 15
The Ticket to Ride Pattern: Regional Administration

This chapter presents a mechanism that assists with regional administration in an internetwork. Using this mechanism, an application messages may be generated with one set of administrative policy and have that policy altered to adhere to the policy of one or more network segments through which the message traverses on its way to a destination.

The ticket-to-ride pattern (TRP) augments messages with specific information describing the policies of the network segment it must traverse. The pattern is named after the concept of purchasing a train ticket. In many rail systems, a ticket, once purchased, must be kept with a passenger while on the train and the ticket may be requested at any time during the journey to prove that a given passenger has paid the correct fee for the ride. If a passenger does not have a ticket to ride, they can be removed from the train just as a message without a policy extension may be rejected from a network segment.

15.1 Pattern Context

Several envisions operational deployments of DTNs span multiple individual networks that have been federated into a single internetwork [1–3]. A message generated on a node in one network (a source network) may be delivered to a node in a different network (a destination network) by traversing one or more nodes in other networks (transport networks). It is the responsibility of a network architect to determine how these individual networks interface with each other, and in certain cases, when a single network should be segmented into one or more independent networks (Figure 15.1).

Network boundaries are created when a node must translate some portion of a message from the format it had upon receipt to the format it must take at transmission. Any node that performs such a translation is considered a boundary node and the portions of the network it joins are considered network segments.

The types of changes that occur across such a network can be characterized as changes to the physical representation of user information, the protocols that format those data, or the policies that describe how those data are handled by nodes in the segment.

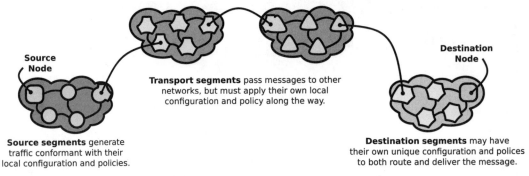

Figure 15.1 Segmented internetworks.

Physical boundaries occur at nodes that transmit a message using a different medium than that used to receive the message. For example, a message received over a Bluetooth connection by a smartphone which then sends the message over a cellular connection acts as a physical boundary bridging two different network segments (Figure 15.2).

Protocol boundaries occur at nodes that reformat a message using a different transport protocol than that used to receive the message. The selection of protocols is sometimes based on the characteristics of the underlying network (such as the selection of a store-and-forward protocol in a challenged network), the availability of support software and hardware (such as when using IP based on its widespread adoption), or the availability of special protocol features (such as extension headers). This boundary occurs when bridging very different types of networks, such as the difference between spacecraft ground stations using the TCP/IP protocols and spacecraft that use CCSDS protocols.

Policy boundaries occur at nodes that add or remove meta-data to messages to affect how the message will be handled by nodes once they leave the that node. These metadata may involve the application of security services (to include authentication), the setting of new report-to addresses, or requiring any other additions, replacements, or removals of annotative data stored in packet headers. For example, one network segment may not require data encryption for messages, but another network segment may require that all data be encrypted. When traversing this boundary, encryption may need to be added (or removed) from a message.

Crossing physical and protocol boundaries require additional processing as part of the translation process. Once the translation has been completed, the handling of the message through the segment proceeds without modification. However, crossing a policy boundary will have longer-lasting effects for the message. Therefore, special care must be taken with policy changes as they may have the most significant impact on overall message delivery and consumption of networking resources.

The TRP provides a mechanism by which a node places additional, annotative information with a message to help provide consistent handling of the message across the variety of network segments it may traverse. This pattern relies on transport protocols that permit inserting extension headers into messages as they journey through a network. As a message enters a network segment, a header specific to that network segment can be included with the message that can later be

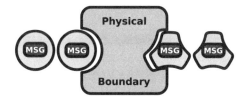

Physical boundary nodes translate the medium used for message representation, but the messages themselves remain unaltered.

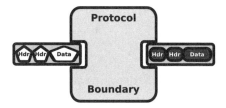

Protocol boundary nodes change the semantic representation of user data through different data formatting and different annotative information.

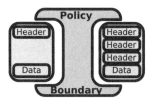

Policy boundary nodes change the information associated with user data to effect how it is communicated and handled through the network segment.

Figure 15.2 Types of boundaries across a network.

removed when the message leaves the segment (or kept for record keeping at the message destination).

Applying this concept to network segments, a message can be issued a ticket to ride on the network segment (see Figure 15.3) where such a ticket can be used to (a) assert the correctness of the message to use network segment resources, (b) hold any changes specific to the handling of the message through the segment, and (c) optionally carry any information that may have been lost when the message first entered the network segment.

15.2 The Problem Being Solved

Challenged networks can benefit from different physical layer technologies, transport capabilities, and policies to use a diversity of technologies for communicating user data in difficult circumstances. To create the best mix of such technologies, messages in such networks will cross through a variety of network segments.

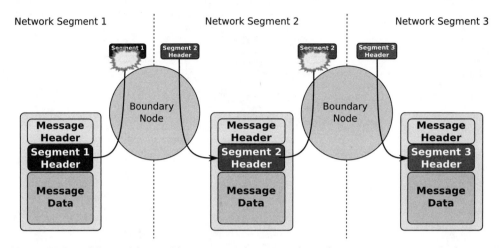

Figure 15.3 Adding a ticket to ride at a network segment boundary.

Protocols such as BP require new features, such as store-and-forward, and avoid traditional features, such as end-to-end sessions, to handle delays and disruptions across these segments.

Because session information cannot be presumed to exist at endpoints, individual messages must carry their own information with them to aid downstream nodes in the network with their processing. For this reason, protocols such as the BP allow nodes to add additional headers that provide the information necessary to extend the default processing of the bundle in some way.

Annotative headers complicate how individual messages traverse different networking segments. If a message embeds policy information in an extension header, and that embedded policy conflicts with policy encoded at a network node, there can be ambiguity on how to process the bundle: some determination must be made as to whether the message policy or the network policy should be used. This conflict must be resolved for every network segment through which a message will traverse. There are five general areas where packet handling may be overridden by a network segment: security services, processing flags, error handling, priority, and resource utilization. Here are some examples of potential conflicts:

- *Security.* If a message contains security-related extensions, such as carrying a specific integrity signature or ciphertext, an accepting network segment might not be able to add its own security results. If a message contains multiple integrity signatures from other network segments there may be confusion over how those signatures are to be verified in the current segment. Further, if the accepting segment requires encrypting data there may be ambiguities as to whether existing integrity signatures are calculated over plain text or cipher text. If the accepting network has a policy that all integrity signatures on data are verified at every hop in the network (to avoid persisting corrupted or malicious data) the accepting network may not have the keys necessary to verify that signature.

- *Processing Flags.* Transport protocols often define flags in headers that describe how a message should be treated in various circumstances. For example, flags can be defined that control whether a message should be fragmented, whether extension headers should be removed if they are not processable by a node, and whether the entire message should be removed if any part of it is unable to be processed. A message source sets these flags with the desired behavior from the point of view of the originating application (or originating segment). However, the policies of the accepting segment may require different behavior. In this case, the accepting network would either disregard its policy or overwrite the policy of the originating or application.

- *Error Reporting.* Error reporting is a common function of an established end-to-end session. Reports are useful so that missing traffic can be NACK, sequence numbers for ACKs can be used to calculate missing data, and timeout values can be used to infer loss of connectivity. In DTNs, these reporting mechanisms may be co-located in the message itself. For example, BP bundles support the concept of a report-to endpoint where status reports can be sent if an error is encountered. This report-to endpoint can be specified by the originator of a bundle and used to collect important metrics associated with the bundle—from the point of view of the originating application or originating network segment. However, the accepting network segment may desire its own report-to endpoints to track the progression of traffic through the nodes for which it is administratively responsible. The accepting network segment, as a function of operational security, may not wish to send details about its internal message traffic to random endpoints outside of its control. Like processing flags, the accepting network must decide as to whether error handling information, if present in a bundle, is to be overridden or not. If overridden, then the current network segment can manage the bundle as desired, but at the cost of removing error reporting requirements placed on the bundle by the originating application.

- *Priority.* Priority in this case refers to the order in which messages are selected for transmission over egress points on a node. Messages with higher priority are selected over messages with lower priority. Setting message priority is as much a matter of message existence as timeliness because low priority messages may constantly have their transmission delayed until they expire from the network. The importance of priority often causes individual network segments to strictly control what messages are given what priorities and these settings are often independent of the priorities bestowed upon a message from some other network segment. For example, a message may be given a priority consistent with the originating network's willingness to commit resources for the message. Some accepting networking might map the incoming message's priority into its own priority set. In doing so, high priority messages from low-priority network segments can be deferred for lower priority messages coming from higher priority network segments.

- *Resource utilization.* Transport protocols often embed threshold or limiters associated with how many resources a message can consume in a given network. For example, in IP there is a limit on the number of hops a message may take in a network, which is an attempt to limit the amount of energy and bandwidth a message can consume (especially if a message is trapped in a routing loop). In BP, there is a notion of time to live that reflects the relative number of seconds of existence before a bundle should be removed from a network. Messages with large hop counts or times to live can occupy storage at nodes for longer periods of time and cause nodes to spend more time transmitting (or retransmitting) data. Similar to priority, an originating application might set very high thresholds for metrics that impact resource utilization because that segment is willing to expend resources on a particular type of data. However, an accepting network segment may not agree to commit such resources and may wish to override these thresholds.

Each of these conditions present practical problems encountered when messages cross network segments in complex and heterogeneously managed networks. There are typically two methods that can be used to address these issues: encapsulation and extension.

By using encapsulation, a network segment provides a clean separation of policy areas. The original message can be expressed as the payload an encapsulating message, and the encapsulating message can be populated with all of the policy settings specific for the network segment. While elegant, there are three problems with this approach.

First, the encapsulation may not preserve vital processing information associated with the original message. For example, if a message were generated with a very long time-to-live because the data it represents is important or hard to duplicate, and the encapsulating message sets the time-to-live to be very short, then the data may be destroyed by the network segment. If report-to endpoints have also been encapsulated, then the originating application may never receive an indication that the message was deleted, leading to the same situations happening again at a later time (either with another message or as part of an attempt to retransmit the original message).

Second, if all data is encapsulated through each new network segment it becomes impossible to perform any deep-message inspection on information at rest in the system. For example, using store-and-forward techniques to congregate data at strategic points in a network is a way to implement certain types of ICN. Inspecting data in a store-and-forward cache to consider opportunities for data fusion and data filtering become impossible if the data is encapsulated.

Third, some messages in a challenge network accumulate data as they persist through the network. For example, extension headers may collect information for nodes visited along a path as a way of disseminating congestion information. In cases where messages use variants of source-path routing, the inability to see the extension headers of an encapsulated message prevent these types of control algorithms from running for an individual message.

For the cases where encapsulation is not an acceptable solution to differing policies, the TRP provides an alternative mechanism.

15.3 Pattern Overview

The TRP describes the logical components and their associated behaviors for implementing extension headers in a message to implement local network segment policy. Using the TRP, a segment can place whatever policy it needs in the extension header and prioritize that information over other information in the bundle. This approach avoids the pitfalls of encapsulation by preserving all original message information in a way that is inspectable (and updatable) at all nodes in the network. The pattern is comprised of three components (see Figure 15.4): a network segment extension (NSE), a segment extension editor (SEE), and an extension policy engine (EPE).

15.3.1 Role Definitions

The NSE is an extension header added to a message when it enters a segment that persists for the time that the message exists in that segment. The NSE contains all segment-specific policy information necessary to manage the message while it exists in the network segment. The NSE is added to a message when it enters a segment and is removed when it exits the segment. At the time of creation, the extension may be populated with the best information available to the ingress node, but this information may evolve over time. Therefore, this extension may be updated by any segment node to reflect changing updated information.

The SEE resides on the ingress nodes into a network segment and is responsible for the construction of NSEs for messages coming into the segment from other segments. The SEE is also responsible for providing an integrity, and if applicable, confidentiality service for the extensions that it creates.

The EPE resides on all nodes in a network segment and has dual responsibilities. First, EPEs instruct the SEE as to what policy must exist in the NSEs that are added to new messages. Second, EPEs evaluate existing NSEs embedded in incoming messages to determine whether any network action is required as a function of processing the message and whether there are any updates necessary to the extension. The EPE must determine what policy and configuration information existing in the message must be overridden by inserting alternate information into an NSE. To do this, it must balance the needs of the network segment with the original

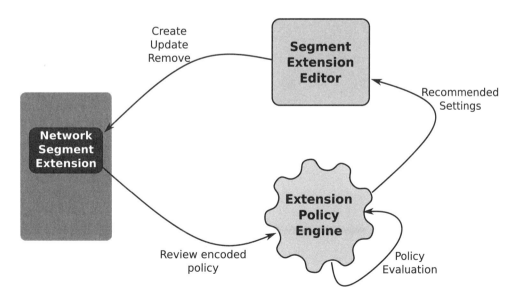

Figure 15.4 TRP component overview.

guidance (if any) present in the incoming message. This includes determining the time-to-live, hop count, priority, security policy, report-to addresses, and other potential differences between the message originator and the current segment.

15.3.2 Control and Data Flows

Data flows through the TRP are focused on the NSE. The SEE creates, updates, or removes the NSE from a message and the EPE reads the extension and recommends changes to the SEE. The complexity of this pattern is in the policy evaluation to determine which aspects of the extension should be present in the NSE. These determinations, as provided by the EPE are network dependent (Figure 15.5).

There are three interesting use cases to consider when determining how components in the pattern interact: ingressing a new message into a segment, routing a message within the segment, and egressing a message from the segment.

When ingressing a message into a segment, the node receiving the message will pass relevant portions of the message to the EPE that will make a policy decision related to how the message should be handled in the network. In this case, the message would either (a) have no NSE attached to it or (b) have an NSE from some other segment attached. In either case, the EPE would determine what policy should be present in the message while it exists in the current segment. That policy would be sent to the SEE which would generate the NSE for the message, apply necessary security considerations (such as integrity and confidentiality) and the add the extension header to the message.

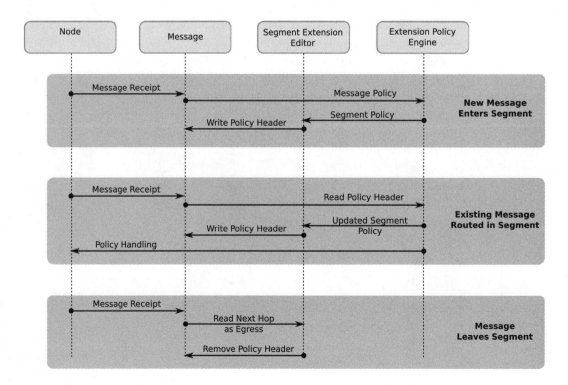

Figure 15.5 Control flow through the TRP.

When routing a message through a segment, each node that receives a message should inspect the segment's NSE by sending it to the EPE for verification. If the NSE must be updated based on the recommendation for the EPE, that recommendation will be passed back to the SEE, which can update the NSE, reapply any necessary integrity and confidentiality, and replace the original NSE in the message. Additionally, any policy updates may also need to be communicated directly back to the receiving node to ensure that it handles the message appropriately.

When a message is about to egress from the segment, the NSE may be removed. In this case, if the SEE if made aware that the next hop for the message is out-of-segment, it can remove the NSE prior to its next transmission. Nodes may wish to remove NSEs at egress points to both reduce the overall magnitude of traffic through the next network segment or to prevent other segments from reviewing internal policies associated with the segment.

15.4 Service Types

The mapping of pattern components to service layers is driven by the most logical layers of the service stack to instantiate data repositories (see Table 15.1).

The TRP exists to help network segments enforce policy and configurations without overriding information captured in a message itself. As such, this pattern is only applicable for those service types that have network-segment scope. This is solely in the area of federating applications: a federating application is one which handles the translation between networks (or network segments).

Transport layers are hop-to-hop making a single node both an ingress and egress point, and therefore, do not need to add any message information to help apply consistent updated between an ingress and egress point. Network and overlay layers, by definition, are homogenous end-to-end and therefore do not support the concept of segmentation as defined by this pattern. Application layers are agnostic of networks and network segmentation.

15.5 When and How to Integrate

This section describes when the TRP should and should not be used and how to make design decisions on the mapping of pattern roles to software applications.

Table 15.1 Check Point Service TRP

Service	NSE	SEE	EPE
I. Transport			
II. Networking			
III. Federating	X	X	X
IV. Overlay			
V. User Endpoint			

15.5.1 When to Use This Pattern

The premise for the TRP is that a message may be pre-encoded with policy or configurations that are incompatible with the local policies of one or more network segments that the message must traverse. The TRP should be considered when all the following conditions are met.

Messages cross network segments that have conflicting policies and/or configurations. If a message never crosses some type of network segment boundary then there is not a conflict in the policy, configuration, or administration of the message. Similarly, if a message does cross segments, but those segments have no differences in how they treat traffic, there is also no possibility for conflict. The TRP should only be used when a message traverses one or more network segments that have conflicting or otherwise incompatible policies and configurations.

Transport protocols encode policy and/or configuration. A network segment must be able to understand the policy and configuration associated with the message. This can only be accomplished if the message itself encodes that information in a standard way. Because transport protocols are responsible for the representation of user and control information associated with message exchange, the transport protocol must include policy and configuration for the TRP to be relevant.

Transport protocols support extensibility through secondary headers. The TRP works by augmenting individual messages with information relative to the current network segment to avoid overwriting information already present in the message placed by the message originator. This can only be accomplished if the transport protocol carrying the message supports some mechanism for adding additional annotative information.

Encapsulation cannot be used across the network segment. Encapsulation provides an elegant way to assert network segment specific information in the primary and secondary headers of a protocol. However, there are multiple cases where the encapsulation of a message prevents the proper processing of that message through the network. The TRP should only be considered in cases where encapsulation is not appropriate.

15.5.2 Recommended Design Decisions

The detailed design of components, and the way in which they should be integrating into a functioning network, are specific to the eventual operational environment. Multiple recommendations exist for how to make design decisions that result in a scalable, efficient, and operational implementation.

The NSE should be signed. The NSE provides control information that determine how a given message can access network resources. The control of network resources equates to the control of the network itself, and therefore is often the target of network attacks. The SEE should sign the NSE with some key that allows EPEs in the segment to both (a) understand that the NSE was not changed since it was last modified and (b) ensure that the NSE was populated by a trusted member of the network segment.

The NSE should be encrypted when there is egress ambiguity. Because the NSE contains information that is only of interest to the segment itself, it should be encrypted such that only nodes in the segment can review the information.

Additionally, to the extent that the NSE may encode addresses and policies that are considered private by the network segment, every effort should be taken to prevent that information from leaving the segment. Encryption may not be necessary if a message will never leave a network segment through anything other than an identified egress node. However, if a message may find itself in a different network segment (because of multicast transmissions, topology changes, or other considerations) then NSE should not be viewable by nodes outside of the segment.

The SEE should preserve existential information for the message. When a network segment overrides control information in a message, care must be taken to understand how those modifications can change the retention of data in the network. For example, if timeout, retention, acknowledgement, and other information is asserted in an original message, a network segment should preserve those settings that could otherwise result in significant data loss.

The EPE should check the NSE at every node. As the capture of policy and configuration for a message, the NSE describes how every node in the segment should process the message. Therefore, every node in the network should check for an NSE in each message and handle the message in accordance with the contents of that extension.

Where possible NSEs should reference policy. Every message through a segment may have an NSE attached to it. If multiple messages flow through a segment with relatively similar message policies and similar segment-specific policy overrides, then the traffic through the segment can be reduced by predetermining a finite set of policy directives and placing a policy reference in the NSE rather than a verbose description of the policy. For example, an NSE may contain simply a policy associated id that would be conceptually like the concept of a security association ID as used in IPSec.

A network segment should remove messages where the NSE is missing. The NSE should be considered the ticket to ride for a message in a network segment. Even in cases where there is no policy or configuration override, the existence of an NSE in a message is evidence that the message went through a proper ingress point into the segment and that policy has been considered. Lack of such an extension should be a message which exists in the segment that, otherwise, should not. Messages sourced external from the network segment and lacking an NSE should be discarded or otherwise quarantined.

15.6 What Can Go Wrong

There are very few guarantees in highly challenged networks, making the deployment of this pattern susceptible to a variety of potential issues.

Radically different policies can break end-to-end data exchange. Changing control-plane information such as time-to-live, hop-count, acknowledgement schemes, and other items can significantly impair control loops used to provide reliability between messaging endpoints. Even challenged network scenarios make use of the concept of timeout-based retransmissions. If a long timeout is provided by a message originator and a network segment significantly reduces that timeout and expires the message prematurely, a message may never fully get through a network segment.

Attempting to mix and match policies can lead to security concerns. Network policies and configurations often work together to provide a holistic view of message management in a network. Attempting to create hybrid versions of policies can dramatically reduce the operational security of the network. For example, if an originating network requests status reports be sent whenever a message is forwarded, and that policy can persist in a network segment, then the network segment will provide progress status reports to an entity potentially outside of that network segment. This could result in an external entity understanding the topology, mobility, and even priority of traffic within the network segment. For very private network segments, this could result in a security breach.

NSEs can increase the size of messages with potentially repetitive data. If policy and configuration overrides for incoming messages are the same, then the same NSE might be added to multiple messages into the network segment. If the NSE is large relative to the incoming messages, this can lead to wasting network resources be repeating large amounts of similar information for messages. For example, if a stream of packets representing a multimedia broadcast traverse a network segment placing a large NSE on each packet could degrade network performance and incur needless processing. In such a case, NSEs should be kept as small as possible, perhaps deciding to only encode a policy association.

15.7 Pros and Cons

This section summarizes the benefits and pitfalls of using this pattern (see Table 15.2).

15.8 Case Studies

The TRP is most applicable where messages must pass through multiple, independent administrative domains. This condition occurs regularly in federated DTNs where each constituent network may have its own policies based on its own unique capabilities and constraints. For the purposes of explanation, a deep-space federated network can be imagined and the TRP discussed in the context of that internetwork.

Table 15.2 TRP Pros and Cons

Design Goals	*Pros*	*Cons*
Preserve Segment Policy and Configurations	Every message in the network can be tagged with specific segment information.	Segment extensions can expose specifics of network administration. Similar NSEs in multiple messages can be wasteful in the network.
Allow Message Processing in the Segment.	By not encapsulating traffic, all original message data is available in the segment.	Segment policies can override the time messages stay in the segment.

15.8.1 Federated Deep Space Networking

Consider the case of a university payload (such as a camera) that has been manifested on a planetary lander build by a national space agency. This lander may use an existing planetary orbiter for backhaul communications that has been built (and operated) by a different national space agency for the purpose of providing a communications network. Finally, consider the existence of a commercially operated space relay network service that has been constructed for the purpose of capturing planetary data [4]. When the university payload generates some data (such as a high definition image), a message is created in which message headers may include some policy information and the message payload includes the generated data. As this message traverses various administrative domains, the policy for how the message is handled will be different.

The journey of this message is captured in Figure 15.6. Within the payload itself, a message is generated and header information may include security, reporting, priority, and other information. However, this message is one of several messages from the particular instrument and the instrument may be one of several on the lander itself. When the message is accepted by the lander it may be issued a "ticket" (Ticket 1) that describes how the lander will treat the information. For example, if the lander must drop the message due to exhaustion of lander resources, then the lander operations center will need to be informed of a problem with the lander itself. This may happen in addition to the payload operations center being informed of the loss of a particular set of instrument data.

When the message is sent from the lander to the orbiter, another ticket may be added to the message (Ticket 2). This second ticket may be necessary if this other space agency's orbiter provides a different data priority, security services, or reporting information for data through the orbiter. The lander in this example may be

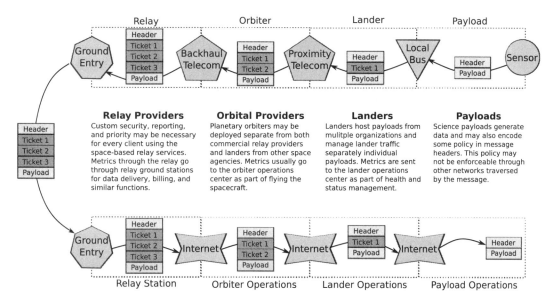

Figure 15.6 The use of TRP in a federated network.

one of several landers serviced by the orbiter and the orbiter may assign priorities based on service agreements that differ from lander to lander. Similarly, when the message is accepted at a relay node, another ticket (Ticket 3) may be necessary to associate the message with some service level agreements for billing. Once the message reaches the ground and is processed by different hops and tickets can be removed as they exit various segments of the internetwork.

Using tickets as annotative information in this scenario yields the following benefits:

- When changing physical layers or protocol layers, information such as the addressing schemes, timestamps, and delays may be included to provide future destinations with source information.
- When changing configurations, information such as new policy configurations, report-to addresses, and processing flags may be added to a message by either (a) representing new policy information for the message in an annotative header or (b) replacing existing policy information with new segment information and storing the original information in an annotative header.
- When changing administrative information annotative information can be used to provide statistics or other information from the older network segment carried into the newer network segment to inform metrics collection and message handling.

15.9 Summary

Challenged networks cannot establish and maintain end-to-end sessions to hold pre-negotiated assumptions between a message source and destination. Lacking the ability to support sessions, messages must carry with them whatever processing information is necessary to help with message forwarding and delivery. Transport protocols, such as the BP, support extension headers that can be added, modified, and removed within the network. These extension headers can be used to carry session-like information.

Often, challenged networks are built by federating multiple individual networks that use different physical media, protocols, or policies. This can be a matter of financial practicality through the reuse of existing networks or a matter of technical practicality where different technologies provide needed capabilities through sections of the network. In cases where differing segments use differing policies, the practice of encoding policy with a message can cause ambiguity on how a message is to be processed by the network: in what cases should message policy be followed and in which cases should segment policy be followed.

One way to handle this ambiguity is through message encapsulation, where a message coming into a network segment is completely encapsulated by another message. In this way, policy decisions can be kept with messages: the encapsulating message is configured with segment-specific policy and the encapsulated message keeps its configured, also encapsulated policy. While this is a feasible solution to the policy problem, it also makes the encapsulated message opaque to nodes within the segment and prevents certain data fusion, data filtering, and advanced routing concepts that are often considered useful in challenged network scenarios.

An alternative to encapsulation is to add a segment policy extension header to a message that can carry segment-specific information. Nodes in a specific segment can choose to process policy from the segment-specific header rather than other headers in the message. This provides the ability to override message-specific policies without cause all parts of the message to be opaque to node sin the segment. The TRP provides a standard way to reason about segment policy headers to provide some consistency in policy enforcement in segmented internetworks.

15.10 Problems

15.1 Provide one example each of a physical, protocol, and policy boundary that exists on the terrestrial internet. Explain why each of these boundaries exists and how nodes determine how to perform necessary translations.

15.2 Give an example of a security processing ambiguity that can be present if there are different versions of security headers in a single message.

15.3 Give an example of different processing flag default settings in two different network segments. Explain why these segments may choose to have different settings for this flag. Describe a consequence if default flags from one segment are used in another segment.

15.4 Consider a message that encodes in an extension header a report-to address to send status reports based on when the message is deleted. In this case, the message source sets this address and in a different network segment message forwarding fails and a status report needs to be sent. Provide two reasons why the report-to address should be kept the same as set by the message source. Provide two reasons why the report-to address should be updated to an address in the current network segment.

15.5 Provide three reasons why a message priority should be changed from its original priority setting set by the message source.

15.6 The text provides three reasons why encapsulation should not be used as a policy override mechanism across subnetworks. For each reason, give a detailed example of how encapsulation results in failure to process data.

15.7 Given three examples of how a message-specific policy can differ from a segment-specific policy.

15.8 Discuss how an EPE can determine which policies encoded in a message should be overridden for a particular segment. Given an example of how mixing message and segment policies can result in unintended message processing.

15.9 Discuss the security implications of having an NSE for one segment be accessible to nodes outside of the segment that created it.

15.10 Describe a mechanism other than using policy associations to reduce the overall data volume increase caused by using NSEs in a segment.

References

[1] Lluch, I., et. al., "Simulating a Proactive Ad-Hoc Network Protocol for Federated Satellite Systems," *2015 IEEE Aerospace Conference*, IEEE, 2015.

[2] Xua, R., et. al., "IoT Architecture Design for 6LoWPAN Enabled Federated Sensor Network," *Proceeding of the 11th World Congress on Intelligent Control and Automation, IEEE*, 2014.

[3] Pollner, N., et. al., "An Overlay Network for Integration of WSNs In Federated Stream-Processing Environments," *2011 The 10th IFIP Annual Mediterranean Ad Hoc Networking Workshop*, IEEE, 2011.

[4] De Sanctis, M., et. al., "Space System Architectures for Interplanetary Internet," *2010 IEEE Aerospace Conference, IEEE*, 2010.

CHAPTER 16

What Can Go Wrong Along the Way: Special Considerations for DTNs

Much of this book has dealt with the benefits of the BP, its configuration, management, and use in network architectures. We have considered cases and scenarios where DTNs provide real advantage over traditional networking architectures. While these advantages are quite real, network engineers and network architects must be aware of cases when DTNs are challenged, constrained, or otherwise poor solutions to a problem. This chapter will identify and address several of the common pitfalls associated with the design and operation of DTNs.

16.1 Resource Limitations

DTNs exist in the real world and as such are affected by real world limitations. Resources in and available to real world networks (as opposed to thought-problem network designs) are limited. Processors only have so many available cycles, nodes have a fixed amount of available storage, and links are fundamentally bandwidth-delay limited. What this means in practice is that while a DTN can be designed and shown to meet requirements and work using idealized protocol behaviors, resource limitations may cause the network to fail once the network is implemented in actual systems.

In practical DTNs, resources come in three flavors: link bandwidth, node processing capacity, and node storage volume. These resources are the same whether we are discussing DTNs or traditional intolerant networks.

- *Bandwidth:* Bandwidth describes the amount of information that can flow from one node to another at any given time. It is an instantaneous measurement. Bandwidth-delay is a way of measuring how much information can flow in a unit of time. The postal service has nearly infinite bandwidth, but long delay. Conversely, cellular telephone connections have limited bandwidth but very short delays. This is why radio telescopes mail physical hard drives with hundreds of terabytes of information to scientists for processing rather than sending it over the Internet (enormous bandwidth and delay is not a factor) [1]. In the other extreme text messages and email have practically killed the paper letter (small bandwidths are needed but a very short tolerance for delay).

- *Processing Capacity:* Processing capacity refers to a node's ability to make decisions about what to do with a given bundle, store bundles, retrieve bundles from storage, and generate and interpret protocol control notifications and messages. The measure of processing capacity is typically made in operations per second—a term from software engineering—representing how many actions can be accomplished in a given time. In addition to processing the bundle traffic, a node's processor must also perform all of the background functions of the protocol and the BPA including communicating with the convergence layers below and application layer above, maintaining routes and network maps, providing time and security functions, performing housekeeping functions such as cleaning up unused storage and discarding expired bundles, and so forth. Finally, a node's processor may be a shared resource of an actual system in which the BPA is only one of many processes on the host.
- *Storage Capacity:* Storage capacity defines how much information can be at rest on a node at any given moment. Storage is used for bundles that are waiting for an available connection to the next node, available bandwidth to transfer the bundle, or available processing capacity for the node to make a routing/handling decision or perform some other operation. Storage is generally physically limited in practical BPA node implementations. Storage capacity is a key design variable in DTNs as the store-and-forward behavior that defines DTNs is absolutely rooted in the node's capability to store information.

If the bandwidth out of a node is less than the bandwidth into the node, the potential exists to have more bundles coming in than leaving. If this happens storage must be used to buffer the traffic. Since some of the bundles on or entering the node will not be immediately forwarded, some of the node's storage must be used to keep them until their next hop is available. As storage fills with traffic, eventually the BPA will need to begin making priority decisions about which bundles to keep and which to drop. Assuming the node has been configured to send notifications, notices of dropped bundles must then be generated and sent back up the line creating additional network load [2].

Processing capacity can be a limiting factor when more bundles are passing through the node or need to have routing decisions made than the processor can handle. In this case bundles will likely be lost on arrival as the processor does not have enough capacity to handle everything. With the processor overloaded, the BPA does not realize that inbound bundles are being missed and so no dropped bundle notifications are sent back to the source, which now believes that the bundle has been successfully forwarded through the network.

There is a complex interplay between processing capacity, storage capacity, and bandwidth in a network, which is really no different in DTNs than in traditional networks. Routers and nodes can both be flooded with traffic to the point that either storage or processing capacity is exceeded. This can happen naturally in the network or as a result of a malicious actor launching a denial of service attack. The lesson to the network designer is that they must have a good understanding of the traffic expected through the network and size links, processing capacity, and node storage appropriately. Tuning the performance and behavior of the BPAs in the DTN is also important to optimize the bandwidth, links, and storage available as well as to ensure that nodes fail gracefully if an overload condition does occur.

16.2 Accepting Reactive Fragments

The concept of reactive fragments allows for nodes in a DTN to attempt the efficient transmission of bundle data when failures have occurred between hops. The scheme is simple in concept but tricky in implementation. Assume that BPA A is transmitting a bundle to BPA B and only part of the bundle makes it across the link. Assume the hop between A->B is using some form of reliable (accountable) delivery protocol, and so A is made aware that the entire bundle was not received at B, but only some part. With reactive fragmentation, node A builds a new bundle containing only that part of the original bundle that failed to cross the hop and forwards the newly created bundle (containing the fragment) to B at its next opportunity. In this way the protocol can move the entire bundle from A->B while limiting the amount of data retransmission required. On the receiving end at node B, the first part of the original bundle is recombined with the second part, reconstituting and the original message. A simply needs to know how to build a new reactive fragment bundle, and B just needs to know how to put the two parts back together in the correct order.

The tricky part comes when the new reactive fragment bundle takes a different path than the original A->B. If another node (BPA C) makes contact with A before A makes contact again with B, and C is a good route to B, then A may elect to forward the partial bundle to B through C rather than waiting for a new direct contact with B. The end-to-end path becomes A->C->B. If the link is marginal, or if multiple links are used as fragments take different paths to reach the destination, it is likely that more reactive fragmentation events will occur. In this case the fragments will eventually make it to the destination node whose BPA must then reassemble the set of fragments back into the proper order and provide any receipt signaling back up the chain of nodes.

While this is an effective way to complete the initial fragmented bundle's transmission, the complexity increases as additional fragmentation happens and additional features are employed in the network.

16.3 Heterogenous Networks

Heterogenous networks refer to networks that bridge or connect several other networks in which protocols, operating systems, service level expectations, and infrastructure are not the same. An example of a heterogenous network might be a corporate network bridging a front office environment consisting of Microsoft Windows computers with an engineering environment consisting of Linux machines. Likewise, a heterogenous network could consist of two internal corporate LAN environments that span a WAN environment operated by a second party service provider. In both examples, the underlying behaviors and assumptions of the several network regions are likely quite different. Even when common protocols are employed—say TCP/IP or LTP/BP running throughout—behaviors at the operating system or physical infrastructure levels may differ. Security and administrative rights may vary between networks. Finally, service-level agreements—and service-level expectations—may differ depending on where in the broader heterogenous network a bundle is at any given moment.

What this means for the network designer is that it's not sufficient to simply say that the network is a DTN or speaks BP to ensure that information is able to flow. The concepts of interoperability and cross-support must come into play. For our purpose, the concept of interoperability between networks refers to two networks that while they may be different in terms of implementation, service level, and supported protocol behaviors still have enough commonality to accomplish the function of moving a bundle from source to destination. The concept of cross-support is less concerned with the technical and more concerned with the agreements and relationships between the organizations managing and operating the networks and the agreement between them to work together to move information.

16.4 Dissimilar Implementations

A frequent assumption is that if two elements have both implemented the same protocol or standard then they should be fully compatible and interoperable. Additionally, there is a general assumption that later versions of a protocol provide backward compatibility to earlier versions. Unfortunately, the fact that two (or more) implementations of a standard have been accomplished does not ensure that they are actually interoperable or compatible (and they are not the same thing). Compatibility can be loosely defined as two things (protocols, elements, nodes, implementations, etc.) being able to coexist without adversely affecting the other. Interoperability, however, adds the flavor of mutually agreed behaviors between the two things such that they can accomplish some desired function together. How a BPA is designed, developed, implemented, configured, and deployed all come into play in determining to what extent it can coexist alongside of and interoperate with other BPAs in the same network environment.

What's more, despite the best marketing efforts of standards organizations, the fact that a standard exists at all does not guarantee that any two implementations are compatible or interoperable. Additionally, as is the case between BPv6 and BPv7, revisions to a protocol specification may obsolete or deprecate functions and behaviors or may change the protocol in other ways that are significant enough so as to make the two versions incompatible.

Some organizations spend at least as much time and effort accomplishing interoperability testing as they do in focused development of implementations of protocols. The CCSDS is a good example of this in that the organization does not generally accept a standard for inclusion into the suite of CCSDS recommended protocols until two independent implementations of that standard have been accomplished and shown to be interoperable through extensive testing [3]. Essentially the standard specification is itself tested by showing that two independent implementations are interoperable when the only source of commonality is what is specified in the standard itself. In this manner the CCSDS ensures that a given standard is written to a quality such that two independent groups will likely produce interoperable implementations simply by following the standard. Relative to BP and DTNs this approach was used extensively by the teams developing NASA's BP suites (ION DTN at NASA's JPL, and DTN2 at the Marshall Space Flight Center).

16.4.1 Dissimilar BPA Extensions

While two (or more) BPAs may be written against a common standard and demonstrate both compatibility and interoperability for the required BP behaviors, this does not guarantee that the two implementations are in fact interoperable. As discussed in prior chapters, the set of defined BP extensions (BPEs) define additional behaviors such as accountability and security that while not fundamental to the task of moving a bundle from source to destination provide additional desirable functionality. Because the BPEs are optional in each BPA it is possible (probable) that two network segments will not support the same BPEs. The BP provides guidance as to how to handle these cases by specifying behaviors when the full extension functionality is not provided. This guidance, while helpful and necessary to ensure compatibility between BPAs, does not provide the actual behavior expected. This means that if a DTN architecture assumes a certain extension functionality will be present throughout the end-to-end network, and one or more BPAs in that network do not support the extension, a gap will exist in the architecture that could cause failures as information tries to transit the DTN.

There are several ways in which two network segments can fail to be interoperable. These can be as obvious as the two segments not implementing the same set of protocols or the same set of protocol behaviors and options, or as subtle as two different operating systems processing data differently. If implementations are compatible there can still be cases in which specific optional functions are not supported—for various reasons—in one or both network segments. This is particularly important when selecting BP extensions to provide critical functions like security in a DTN.

16.4.2 Dissimilar Service Level Expectations

Where the prior pitfalls have addressed technical issues with dissimilar protocol implementations or network functions, service level expectations require the network designer or application developer to consider not just the technical factors of implementing the DTN but also the operational and policy factors in which the real network will exist. Service level expectations are those sets of behaviors and performance standards to which a network (or network providing organization such as an ISP) has committed to. Service level expectations can (and often do) include such things as bandwidth allocation and prioritization schemes, storage resource prioritization, traffic shaping approaches, reliability, and time to restore to service after a fault. Additionally, service level expectations can include operational and organizational factors such as the processes by which one organization reaches agreement with another, the relationship between the service provider and the client organizations, how faults are addressed, and what happens when there is too much demand for available resources among several customers.

Just as different network enclaves can be (and likely will be) made up of dissimilar protocol implementations or built on different technology, the various enclaves spanned by a DTN will likely have dissimilar agreed service levels. This creates a difference between the behavior, process, or performance expected by one network element and that which is actually provided by another. In some cases, such as how to handle resource constraints, this can be addressed by protocol behaviors themselves. In others, such as how to report and respond to failures and faults in the infrastructure, this can cause real difficulty to the network and must therefore be accounted for by network designers and operators.

Service level expectations are described by humans rather than protocols or machine implementations. These are the mutually agreed to behaviors, performance, and expectations that a service consumer (customer) can expect from a provider. Several examples of service level assumptions in a DTN architecture and concept of operations are:

- Assuming that a terrestrial network will have a certain high degree of connectivity between BPAs and their underlying transport layers (e.g., routers, cables)
- Assuming a certain time to restore to service following an infrastructure outage.
- Assuming that nodes in a mobile network (e.g., a rental car or delivery fleet) will make contact with the rest of the network (perhaps at a depot or charging station) with some degree of regularity.
- Assuming that communications links will come and go on specified schedules as is the case for space mission design whose communication links are carefully scheduled and managed between ground stations and distant spacecraft.

As the network is being designed and provisioned, these assumptions become agreements on how and to what extent operational factors can be depended on. We agree that the reliability of the underlying network hardware will be good enough to some number of 9s. We agree that the delivery van will return to its depot at least every 24 hours. We agree that a space mission will be allocated so many contacts to a ground station per week. While quite obviously not technical factors, these agreements (and failure to meet them) will have a large impact on the performance and even viability of a given DTN architecture. It is important that the network designer consider not just the technical behaviors and performance of the DTN but also consider the operational characteristics of the various network segments and providers.

16.4.3 Dissimilar Operating Environments

Operating environments take two forms: The micro in which the software implementing the various protocols and behaviors of a DTN operate, and the macro or physical environment in which the node (and whatever its hosted on) must reside.

16.4.3.1 The Micro Environment

Anyone who has tried to port a piece of software from one computer operating system to another has encountered the truth of dissimilar operating environments on the micro scale. The processors, memory and other storage systems, communications hardware and interfaces, operating systems, and parallel processes and applications all play a role in both how the protocols underlying a DTN are implemented and how they behave. Processors and computer architectures handle data arrangement and storage differently. Large data centers designed to stream terabits per second to end users are implemented in very different hardware and software than data loggers on a herd of zebra or spacecraft avionics on a Martian relay orbiter [4, 5]. A video streaming service or search engine may prefetch and forward position large amounts of data to BPAs at the edges of its network. A space relay on the other hand must be stingy in conserving both link bandwidth and onboard resources [6]. A space relay is poorly suited as a node for forward positioning videos and a large data center has no need to overly constrain storage. When the two come together—say to stream a movie to explorers on Mars—very careful attention must be paid by the network designer in how they tune the DTN architecture taking into account the differences in platforms, performance, and resources across the end-to-end network. It is not sufficient to simply launch the bundles in the direction of the next BPA.

16.4.3.2 The Macro Environment

Dissimilarity at the macro environment level likewise must be considered by the network designer. Used here, the macro environment refers to the surroundings—virtual and physical—in which a node resides and operates. Again, this set of factors takes on the feel of a concept of operations in that the network designer must consider how nodes will interact with each other and the world around them.

Our data center (see Figure 16.1) from the previous example has nodes that reside in powerful industrial servers contained in multiple equipment racks. They are physically connected by copper or fiber optic cables capable of carrying enormous amounts of data. The link and physical layers of the network stack are in most cases homogeneous (e.g., Ethernet over Cat-5 cable). The connections are steady and reliable and excepting faults or maintenance events, can be counted on to provide a connection at all times. The nodes in a DTN built on such a system can assume

Figure 16.1 Data center. (Image credit: Google.)

a level of constancy in the network topography and generally must only provide additional rerouting capability to address failure modes. Such a system has no need for functions like neighbor and contact discovery or link establishment, nor does it need to address nondeterministic routing techniques—static routing will be fine.

By contrast, a DTN consisting of mobile nodes, such as data logging collars on animal herds, is a very different beast. In this case the physical environment for the BPA node is a low power, low size processor with a limited amount of storage space (Figure 16.2). Power is of the utmost importance as there isn't much mass for batteries (at least not if you want to keep the zebra happy!). There are no wires connecting the nodes in the herd to each other and the location of any given node will change as the animals move. This network needs a form of wireless communication (such as Bluetooth or 802.11) as well as the ability to find new connections with neighboring nodes as they move in and out of contact—perhaps implementing additional MANET protocols and techniques beyond the core BP functionality [7]. The BPAs in this DTN must be able to not only identify new neighboring nodes but also to replan routes across the network on the fly and so different (dynamic) routing protocols must be used.

16.5 Noah's Data Ark: A Case Study

Both of these cases are perfectly valid implementations of a DTN architecture. They both provide all of the core BP functionality, move bundles around, perform store and forward functions, and so forth. Now bring them together and make them work as one network, flowing bundles both from the data center to the zebras and from the zebras to the data center. How would we go about this?

Bring the herd of zebra—collars and all—into the data center and try to send a bundle from the zebra to a server in the data center. Nothing will happen. There is no physical connection to move a data frame from the zebra's collar to a port on an Ethernet hub. If that is bridged—say by hooking a cable between the hub and a jack on the collar or setting up an 802.11 WAP in the data center—there is still no way for the BPA node in the server to detect the collar and vice versa. The collar node is able to perform neighbor discovery, but the server node doesn't know how to respond to the queries. Likewise, the server node can send a bundle but doesn't

Figure 16.2 ZebraNet. (Image credit: [5].)

have a static route configured to send to the collar node. Both are fully capable of sending and receiving bundles and doing all of the useful functions of BP, but they don't know about each other and so have no way to route from one node to the next.

Again, helping them out, we configure both nodes with a static route between the collar and the server. Bundles can flow. Traffic flowing from the network segment consisting of the collar nodes will probably be just fine getting to and being processed by the server nodes. Traffic in the other direction however—from the servers to the collars—will likely very quickly consume the resources of the collar node serving as the first hop and the flows will bog down. At this point the collar nodes will very politely begin discarding bundles. If the collar node sends notifications to the server node that it is dropping bundles (as it should to provide accountability across the DTN), fairly soon all of the bandwidth out of the collar node BPA will be consumed with these messages and the network will fail due to congestion.

Something needs to be done to tell the server BPAs about the resource limitations of the collar node BPAs. But that something is not inherently part of the BP and must be implemented through configuration, careful tuning of the network, or additional protocols. This is left as an exercise for the network engineer.

16.6 Working within an Overlay Network

While the BP can and does function as a complete network layer, it is designed primarily to operate as an overlay network (see Figure 16.3) built upon other network segments. The implication of this is that while bundle traffic and bundle-aware applications gain the benefits of a DTN, the overall performance of the network is still affected by the behavior and performance of the underlying network segments. Ideally the underlying segments can be thought of as virtual wires of infinite bandwidth and zero delay. Practically they are networks and come with constraints in terms of bandwidth, latency, reliability, and connectedness.

16.6.1 Routing and Name Spaces

A DTN based on BP will have an addressing scheme and name space to identify and distinguish individual nodes and services from each other. The addresses used

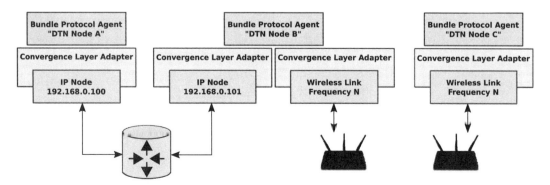

Figure 16.3 Overlay network.

in a given DTN do not translate well into IP or MAC addresses, SONET and ATM switch paths, radio frequencies, or TDMA windows, each of which may be used by one or more of the underlying network segments upon which the DTN overlay network is built. This complicates routing and bundle delivery in that the BPA's routing algorithm needs to have knowledge of not just the DTN's members and topology but also the topology of the underlying network segments.

As an example, a BPA wishing to send a bundle across a TCP/IP link will need to know the IP address of the local node and the IP address of the node hosting the remote BPA in addition to knowing that in order to reach the remote BPA the bundle should be sent across a specific TCP/IP network segment as IP traffic originating at its local IP and destined for the remote IP. The DTN does not need to concern itself with the underlying routing of the IP network, but it does have to have some insight into the expected end points. This only applies hop by hop between BPAs as routing among BPAs across the DTN is handled at the bundle layer.

16.6.2 Networks in Motion

This case of BP over IP is simplified but it provides a good example of what can go wrong (Figure 16.4). In this case the BPAs routing algorithm has recorded that DTN node B can be reached through the IP convergence layer at IP address 192.168.0.101. So long as nothing changes this will work just fine. If, however, there is a change in the underlying IP network—say the network uses dynamic IP address assignment or network address translation (NAT) techniques—then DTN node B may move to a different IP address. If this happens, the route from A->B will no longer be valid and must be deleted from node A's bundle routing algorithm. There are a number of techniques for neighbor discovery and route path verification, but this example serves to show a case where while the DTN is an overlay network and ideally insulates the bundle layer from the underlying network segments, behaviors below the overlay can and do have effects on the DTN.

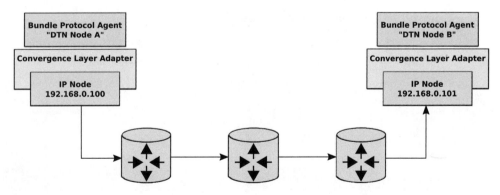

Figure 16.4 BP over IP.

16.6.3 What's in a Name?

In order for bundles to travel across a DTN there must be a way to uniquely identify source and destination nodes. There are several ways to do this that are unfortunately implementation dependent. The general approach defined in RFC5050 uses the internet's URI scheme of scheme_name:scheme_specific_port [8]. The DTN2 implementation took this a step further defining a "DTN" URI of the form dtn://machineID/AppID [9]. The IPN scheme (built on BPv6) for naming DTNs conforms to the Compressed Bundle Header Encoding method and takes the form ipn:node_number.service_number [10]. Since there is no governing organization defining name or address spaces for DTNs (yet) two separate networks could both define a dtn://MyDTNNode/MyDTNApp that are actually different nodes. Likewise, one network could address a node as dtn://Node42/App42 while another addresses the same node as ipn:42.42. Both are correct and valid names for the node but are totally incompatible in practice.

For a DTN to function in practice the nodes within the network need to be assigned unique identifiers and those identifiers need to be used consistently across the network. DTN name spaces are very much network-by-network things, as there is no concept yet of a public DTN address space or DTN address registrar such as that provided by IANA for the internet [11].

16.7 A Network Is a Network

This chapter has focused on conditions that can cause DTNs to fail or operate poorly. These conditions also often exist in traditional networking as well. When designing a DTN, the network designer should also consider traditional networking concerns such as traffic prioritization and shaping, traffic types, bandwidth reservation, link, network and physical layer security. At some point, a network—whether disruption-tolerant or -intolerant—is just a network.

16.8 Summary

This chapter has explored several real-world factors that network engineers must consider when designing, configuring, and operating DTNs. Resource constraints in terms of bandwidth, processor, and storage capacities will exist in any real DTN. These constraints cause the need to tune the performance of any actual DTN so as to minimize the likelihood of resource exhaustion.

Interoperability and compatibility must be considered when bringing multiple platforms and implementations of the DTN protocols into a single network.

Finally, routing and name spaces must be considered, and a method provided—whether static or dynamic, manual or automated—to assign, manage, and update routes as connectedness changes.

16.9 Problems

16.1 Describe how the node resources of bandwidth, processing capability, and node storage volume are interrelated. Provide three examples of how changed to one of these resources affects the other two.

16.2 Design a protocol to negotiate bundle rates between two BPAs with different processing capacity. Assume that the protocol control messages are sent in band as bundles themselves.

16.3 Design a protocol to perform neighbor discovery. What functions must it perform? Where does it need to sit in the network stack? How does it communicate to the BPA and the link layer?

16.4 Provide an argument for why reactive fragmentation can increase the goodput in a network. Defend this argument with an example scenario.

16.5 Provide an argument for why reactive fragmentation can reduce the goodput in a network. Defend this argument with an example scenario.

16.6 Explain how different BP extensions may cause different behaviors in different networks. Provide an example.

16.7 Can a BP overlay completely insulate traffic from different service level implementations? Why or why not?

16.8 If different BPAs are implemented in different operating environments will they handle bundles differently if they otherwise support the same bundle versions and bundle extensions? Why or why not?

References

[1] "What Are Radio Telescopes?" National Radio Astronomy Observatory, August 2019, public.nrao.edu/telescopes/radio-telescopes/.

[2] Cerf, V., et. al., "Delay-Tolerant Network Architecture," *IETF RFC 4838*, Informational, April 2007, http://www.ietf.org/rfc/rfc4838.txt.

[3] *CCSDS Organization and Processes for the Consultative Committee for Space Data Systems*, CCSDS Record (Yellow Book), CCSDS A02.1-Y-4, Washington, DC, CCSDS, April 2014.

[4] "Data Center Innovation|Google Cloud," Google, cloud.google.com/about/data-centers.

[5] Zhang, P., et. al., "Hardware Design Experiences in ZebraNet," *Proceedings of the 2nd International Conference on Embedded Networked Sensor Systems—SenSys*, November 2004, doi:10.1145/1031495.1031522.

[6] Edwards, C. D., et. al., "Proximity Link Design and Performance Options for a Mars Areostationary Relay Satellite," *2016 IEEE Aerospace Conference*, 2016, doi:10.1109/aero.2016.7500680.

[7] Ott, J., et. al., "Integrating DTN and MANET Routing," *Proceedings of the 2006 SIGCOMM Workshop on Challenged Networks—CHANTS*, September 2006, doi:10.1145/1162654.1162659.

[8] Scott, K., and S. Burleigh, "Bundle Protocol Specification," RFC 5050, DOI 10.17487/RFC5050, November 2007, https://www.rfc-editor.org/info/rfc5050.

[9] Fall, K., Burleigh, S., Doria, A., Ott, J., and Young, D., "The DTN URI Scheme", Work in Progress, draft-irtf-dtnrg-dtn-uri-scheme-00, September 2009.

[10] Burleigh, S., "Compressed Bundle Header Encoding (CBHE)," Work in Progress, draft-irtf-dtnrg-cbhe-09, September 2011.

[11] "Internet Assigned Numbers Authority," Internet Assigned Numbers Authority, www.iana.org/.

CHAPTER 17

The Solar System Internet: A Case Study for Delay-Tolerant Applications

Previous chapters have discussed the nature of challenged networking environments, the motivation for delay-tolerant networking, and patterns for building delay-tolerance into networking architectures and applications. This chapter provides a case study in the development of a SSI as a practical approach to adopting a networking layer for deep-space communications. This study provides one example of how delay-tolerant applications and delay-tolerant architectures are being used or considered for use to address real-world challenges. While focused on deep space, the SSI concept manifests challenges that are similar to those encountered by a variety of emerging, challenged terrestrial networks such as remote sensing, vehicular networks, ICN, and high-bandwidth links.

17.1 Overview

The SSI is a concept to add a networking layer to deep-space communications (Figure 17.1). An SSI would provide reliable routing of data across a diverse set of physical links, and in doing so, increase the resiliency and throughput of space communications. Instantiating an SSI would also allow certain types of spacecraft to be built with less expense and complexity by not having to design direct-to-Earth (DTE) communications.

Much of NASA's (and other space agencies') communications concepts now rely on the idea of routing information across networks of heterogenous communication links between Earth ground stations, interplanetary spacecraft, planetary orbiters and landers, and dedicated relay spacecraft.

The SSI Operations Concept [1] was formally published in 2010 by the IOAG. The IOAG[1] serves as a forum for national space agencies to coordinate space communications policy, procedures, interfaces, and other issues that address data exchange amongst spacecraft. This operations concept document described several motivations for the construction of an SSI, to include Mars exploration, Earth science, and Human exploration. In 2014, the CCSDS published an Architecture Description [2] of the SSI concept defining key features of the SSI, to include satellite handover, data rate buffering, reliable delivery, data prioritization, and resiliency through path diversity.

---

1. https://www.ioag.org/default.aspx

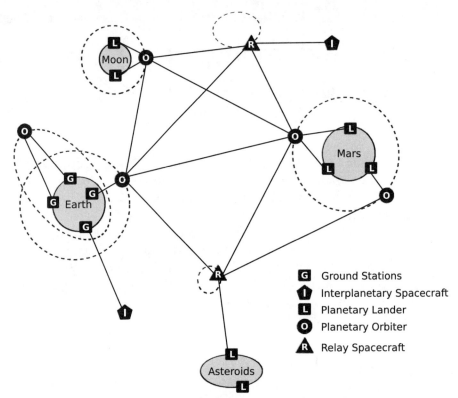

Figure 17.1 An SSI reduces the need for DTE links.

17.2 Motivation

One practical motivation for the SSI is to construct a communications infrastructure at Mars. Earth-Mars alignments make optimum, low delta-v interplanetary trajectories available approximately every two years. NASA, ESA, and other space agencies take advantage of these alignments to plan increasingly complex and ambitious sets of orbiters and landed robotic missions to Mars. As with all deep-space missions, a key performance limiter is the ability to return mission science data from these spacecrafts to Earth. While DTE links can support a mission, opportunities for communication are severely limited by the rotation of Mars relative to Earth and available bandwidth is difficult to provide for landed missions that are extremely constrained in terms of size, weight, and available power.

To address these constraints, NASA, ESA, and Russia developed a concept of operations that involved the use of short range ultrahigh frequency (UHF) radios for communication between the Martian surface and orbit. UHF was selected for its power efficiency, ease of use, and relative ease of accommodating radios and antennas capable of reasonable data rates over distances from hundreds to thousands of kilometers. With this mission concept, a landed spacecraft can communicate DTE when the Earth is in view, but also to communicate using a UHF space-to-space, or proximity link between the surface and a compatible orbiting spacecraft. Since the ground track of the orbiting spacecraft typically offers several contacts between a surface mission and the orbiter per day regardless of the rotation of

Mars this provides many opportunities to relay communications between Earth and the landed mission.

17.3 Experiences and Experiments

A variety of experiments and engineering achievements over several decades have served to characterize the potential benefits of the SSI concept and to demonstrate the proper functioning of enabling networking technologies. This section briefly reviews several significant milestones in the evolution of deep-space networking.

17.3.1 Mars Relay Communications

The early 1990s was a challenging period for Mars exploration and for the development of a Mars communication relay (Figure 17.2). Beginning in the early 1990s, NASA and other space agencies began a long-term campaign of robotic Mars exploration. NASA, ESA, and Russia (then the Soviet Union) planned a series of orbiters, landers, and a set of French-built balloons. NASA's Mars Observer carried a dedicated UHF radio (the MR) to provide relay communication between the French balloons, the Soviet landers (part of the Soviet Mars '92 mission), and the Mars Observer spacecraft. Mars '92 did not launch, and Mars Observer was lost. Fortunately, NASA's Mars Global Surveyor mission (MGS) carried a copy of Mars Observer's relay radio to communicate with the upcoming Russian Mars '96 mission landers. MGS made Mars orbit on September 11, 1997, but the Mars 1996 mission was lost following launch.

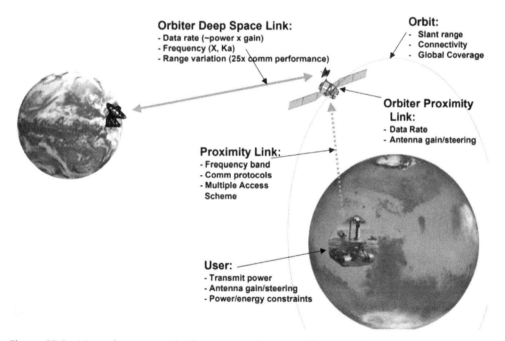

Figure 17.2 Mars relay communication concept. (Image credit: NASA.)

The next batch of missions included the Mars Polar Lander (MPL), the Deep Space 2 microprobes, and the Mars Climate Observer spacecraft. MPL was capable of DTE communication, but the equipment necessary for DTE was too large and too heavy for the DS2 microprobes. As such, DS2's landers were intended to communicate with Earth only through relay—baselining the Mars Climate Observer with its upgraded UHF relay radio for this task. MCO was lost during aerobraking at Mars, and MPL crashed into the Martian surface in December 1999.

The loss of MCO and its dedicated relay payload left MGS as the only remaining Martian relay-capable orbiter until the arrival of Mars Odyssey in 2001. Both MGS and Odyssey carried dedicated relay payloads and were purpose built to include a secondary communications mission objective. While the ability to communicate and measure Doppler rates of other spacecraft was built into both MGS and Odyssey, the planning and execution of relay communication events was manpower intensive. First the relative positions and look angles of the orbiter and the landed spacecraft needed to be analyzed. When a contact period was found in which the lander's antenna was in line of sight of the orbiter's antennas these contacts were carefully planned and scripted so that each spacecraft knew in advance exactly what had to be done to establish the contact, move the data, and then close the contact out.

Since relay protocols had not yet been developed, the actual moving of data required the sender to transmit a file of information as a stream of data encoded as commands to the receiver. The receiving spacecraft would then store the data in local memory to either execute as preplanned command scripts (relay-to-surface) or as a block of mission data to be returned to Earth (surface-to-relay). Only once the data made it back to Earth and was verified as complete and correct could a command then be sent back to the lander (potentially through another relay contact) releasing the held memory for new information.

The Mars Exploration Rovers (MERs) SPIRIT (MER-1) and OPPORTUNITY (MER-2) arrived at Mars in January, 2004. Based on NASA's positive experiences with relay communications the MERs (while capable of DTE) were developed with relay as their primary method of communication [3]. In addition to baselining relay as a first choice, the MERs carried a new CCSDS data link protocol called Proximity-1 [4–7]. Most significant in the implementation of Proximity-1 was the inclusion of the concept of PDUs and the CCSDS Packet Delivery Service, which allowed information to be transferred in packetized chunks rather than as continuous streams of bits. With Proximity-1 much of the burden of planning and executing relay contacts was moved from the flight control team on the ground into an autonomous function in space. This protocol allowed the transition of the Mars Network from a scheduled switchboard model into an autonomously switched system for moving data across multiple hops in deep space (Figure 17.3).

17.3.2 Deep Impact Networking Experiments

The success of Proximity-1 proved the ability to automate communications between two spacecraft relatively close to each other, but the DTN protocol suite was necessary to automate a functioning multi-spacecraft network. Several DTN protocols were tested by the Deep Impact Networking (DINET) experimentation program

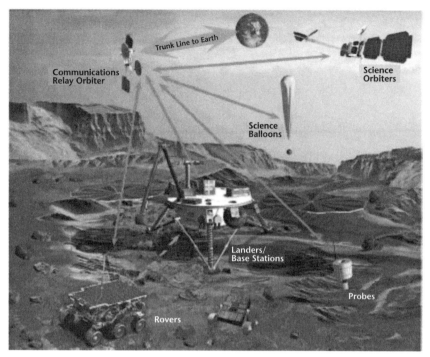

Figure 17.3 The Mars network concept. (Image credit: NASA.)

conducted by the NASA JPL and its partners the Johns Hopkins University Applied Physics Laboratory (JHU/APL), Colorado University, and Ball Aerospace [8–11].

The DINET scenario included simulated Earth, Mars and Phobos locations as well as an in-space relay. The Deep Impact spacecraft was used to simulate a deep-space relay. Nodes representing Earth, Mars, and Phobos existed in various laboratories at JPL in Pasadena, CA. The program consisted of two experiments, DINET-I, which occurred on the Deep Impact spacecraft, and DINET-II, which was performed solely on testbeds.

Figure 17.4 depicts the interplanetary network modeled during the DINET-I experiment. Nodes E1, E2, and E3 represented Earth-based control and payload operations centers connected by traditional networking. Nodes P1 and P2 represented a small network on Phobos, while nodes M1 and M2 represented a similar small network on Mars. Nodes labeled DSN represented various Deep Space Network (or other planetary) communication sites. Finally, the Deep Impact spacecraft acted as the deep-space relay between planets. While the planetary nodes were simulated in computers at JPL, the physical spacecraft was at a distance of approximately 15 million miles (24 million kilometers) from Earth.

The DINET I experiment occurred over a period of approximately four weeks during October and November, 2008 during which time Deep Impact ranged from approximately 49–81 light seconds away from Earth. Eight low rate DSN contacts were accomplished and DINET moved 292 images through the network during the experiment (approximately 14.5 MB). DINET 1 demonstrated the first instance of an interplanetary network on October 20, 2008, when images were successfully received at JPL from the Deep Impact spacecraft located approximately 80 light

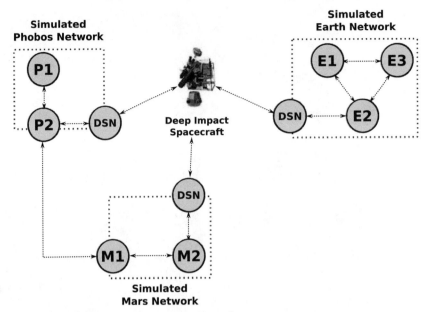

Figure 17.4 The DINET-I experiment network topology.

seconds from Earth. DINET I demonstrated DTN's bundle origination, transmission, acquisition, dynamic route computation, congestion control, prioritization, custody transfer, and automatic retransmission procedures, both on the spacecraft and on the ground.

DINET II (see Figure 17.5) was conducted in November, 2010 using only testbed assets with the objectives to test higher complexity DTN functions including an extended priority approach, implementation of contact graph routing and CGR management, management of network assets themselves, an implementation of a bundle authentication mechanism.

17.3.3 Multi-Purpose End-To-End Robotic Operation Network

The Multi-Purpose End-To-End Robotic Operation Network (METERON) [12] simulates human space exploration scenarios where astronauts orbiting a celestial object remote-control one or more robotic landers on the object. The motivation for remote-control of planetary robots is to avoid the complex, risky task of landing humans on planets and moons without losing the ability to dexterously explore these environments in near-real-time. The METERON concept (see Figure 17.6) is enabled by the concept of a space internetwork as humans orbiting an object would need constant communications with landed assets but will not have constant line-of-site with those assets.

To simulate a realistic test environment, the METERON team devised an architecture that provides actual space links (via NASA's Space Network/TDRSS and Russia's ground network) with the terrestrial internet. A DTN using the BP was established as the end-to-end data architecture between the METERON control center (located on Earth), the in-space robotic control workstation (onboard the ISS), and the rover facility (located at DLR) simulating the robot operating in a simulated Martian environment. This allowed for investigation of long-range

17.3 Experiences and Experiments

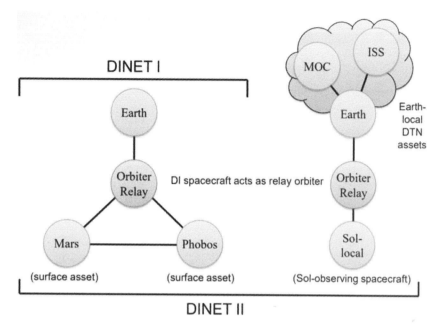

Figure 17.5 DINET-II architecture. (Image credit: S. Burleigh.)

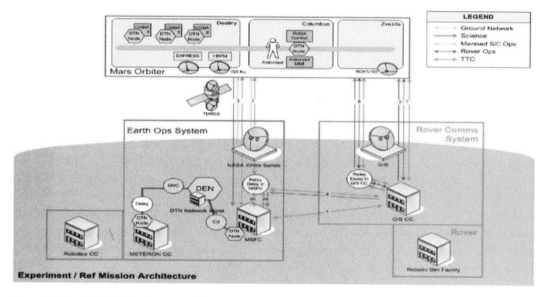

Figure 17.6 METERON reference architecture.

teleoperation between the Earth-based control center and the robot as well as in situ teleoperation from crew located onboard the ISS. Significant in the experiment is the use of actual ISS flight crew operating in a microgravity environment. This allowed the investigators to examine not just the network behaviors and effects on teleoperation, but also to explore the differences that a zero-G, microgravity environment has on human-machine interaction.

An ongoing set of experiments have been conducted since 2013 with increasingly complex mission concepts, robotic systems, and human-machine interfaces. The knowledge gained from METERON has applicability to space exploration, but also to teleoperation and telemedicine on Earth enabling capabilities like robotic support to disaster recoveries and search and rescue, telemedicine/telesurgery, and teleoperation of robots in harsh environments [13] (Figure 17.7).

17.3.4 International Space Station

Beginning in 2009, the ISS Program has taken a series of steps to explore, validate, and deploy BP in a DTN architecture supporting in-space science and ISS operations. Beginning with the BioServe Commercial Generic Bioprocessing Apparatus (CGBA) set of payloads [14], NASA and its partners at BioServe and CU-Boulder began deploying ever more mature versions of the BP and supporting protocol suite (Figure 17.8). Initially intended as a commercially available host for biological experiments in space, the team implemented early versions of BP as an end-to-end DTN between the CGBA payload, the NASA's Huntsville Operations Support Center (HOSC) at the Marshall Space Flight Center (MSFC), and the BioServe and CU-Boulder investigators. As BP and DTN concepts and implementations evolved the CGBA (and its sister payloads) were updated to provide newly developed services and protocol capabilities. The CGBA demonstration of DTN served in large part to validate the concept of DTNs to NASA and reduce the risk of fully implementing DTN architectures to the ISS program.

Through evaluation of BP and DTN's, NASA reached the following conclusions:

1. The use of DTN decreases mission operations labor costs via automated command and telemetry transmission and receipt,

Figure 17.7 METERON SUPVIS-JUSTIN experiment under teleoperation from ISS. (Image Credit: ESA.)

Figure 17.8 NASA astronaut Jim Voss with the BioServe/CU-Boulder CGBA payload on ISS.

2. DTN decreases the need and associated infrastructure and O&M costs for custom control center implementations to interface to nonstandard payloads,
3. DTN decreases the cost of providing communications for spacecraft and payload command and telemetry (C&T) systems by providing a common, standards-based networking protocol and architecture,
4. DTN improves communication link utilization efficiency by minimizing the required data retransmission necessary to ensure successful reception at its destination.

Based on these insights, NASA's Advanced Exploration Systems and International Space Station Programs chose to implement BP and a DTN architecture throughout the ISS payload data network [15–16]. This is significant in that prior to this decision, the ISS payload network (or Payload LAN) was a dedicated, packet switched network based on CCSDS telemetry packets carried over an adapted IP network. The upgrade modified NASA's Telescience Resource Kit (TReK) to include delay-tolerant application support and support for the BP and its associated protocol suite as a native network capability. At the network layer bundles replaced the previous datagrams resulting in a native BP network between systems onboard the ISS to the MSFC HOSC. Once in the HOSC bundles can be sent natively to investigators and science operations centers or encapsulated in IP using a convergence layer and transmitted across the terrestrial internet. This implementation architecture uses the concept of DTN as an overlay network in which the BP bundles span multiple underlying subnetworks, their associated network, and link layers to provide a seamless end-to-end BP environment.

17.4 Significant Challenges

Like any challenged networking environment, deep-space communications are constrained by a variety of factors relating to connectivity, signal propagation delays, and general bandwidth limitations. This section describes the way in which these challenges are unique to the space environment, and thus, difficult to eliminate as a risk.

17.4.1 Connectivity

Orbital mechanics, the position of planets, asteroids, moons, and other bodies, the trajectory and attitude (pointing) of the spacecraft, and other geometrical factors all effect the availability of a communication link at any given time. Geometry-based factors mean that just because an asset is available to communicate with the mission, the mission may not be visible at any given time, and so the scheduling algorithm needs to account for both the traditional concept of network resource availability but also account for the motion of the mission, relays, and celestial bodies as well in order to schedule times when the communication link is physically possible.

17.4.2 Delay

Nodes in the SSE must communicate across astronomical distances. LEO is 400 km away, and the geosynchronous orbit in which most communication relay satellites reside is 36,000 km (nearly the circumference of the Earth). At GEO distances, light takes 120ms to travel directly to and from the surface of the Earth. What this means is that for a one-way relay hop between two locations on Earth, there will be a delay of nearly 0.25 seconds introduced simply due to the time it takes for the signal to cross the distance. This delay is interchangeably referred to as propagation delay, light speed delay, or light speed round trip time (if two way) in communications engineering (Table 17.1).

Node mobility in the SSI is often along parabolic or hyperbolic trajectories curving through 3-D space and the velocities along those paths are constantly changing as a result of gravity. When the distance between two nodes is significant enough, a receiver may have moved beyond the reception area of a sent signal by the time the signal has propagated to the receiver's original location. This transmission problem is conceptually like throwing a ball to someone who is running across your field of view—the thrower cannot aim for where the runner is at the time the ball is thrown but, rather, where the thrower thinks the runner will be. Similarly, SSI nodes must send data to where the receiver will be, not where the receiver is and likewise, the receiver needs to be expecting the signal to arrive from where the sender was, not where the sender is.

17.4.3 Bandwidth

In space communication, the distances are so great, and the available power so limited that bandwidth is fundamentally limited by available power. Space communication links are limited by an inverse square law governing the propagation of energy (RF and optical) across distances. Since the available bandwidth (in terms of data bits per second) is linearly correlated to the signal power (actually the signal to noise ratio) the available data rate also decreases in an inverse square fashion with distance. In practice, when coupled with the SWAP limitation of spacecraft, space

Table 17.1 Example Solar System Propagation Delays

Destination	Distance (in km)	Distance (in AU)	Light Speed Delay (light-seconds)
LEO	400 – 2000 km	$2.7 \times 10^{-6} – 13.5 \times 10^{-5}$ AU	0.001 – 0.005 ls (1 – 5 ms)
GEO	36,000 km	2.4×10^{-4} AU	0.12 ls 120 ms
Moon	384,000 km	2.6×10^{-3} AU	1.3 light seconds
Lunar L2 Point	448,900 km	3×10^{-3} AU	1.5 light seconds
Mars	$54.6 – 401 \times 10^6$ km (225×10^6 km avg)	0.365 – 2.68 AU (1.5 AU avg.)	180s – 1337s (~22 min) 750s (12.5 min) avg.
Jupiter	$588 – 988 \times 10^6$ km (779×10^6 km avg)	3.93 – 6.6 AU 5.2 AU avg.	32.6 – 54.9 min 43.3 min avg.
Pluto	$4.19 – 7.48 \times 10^9$ km	28 AU – 50 AU	232 – 415 min (3.9 – 6.9 hours)

communication link designers are faced with a design space that requires significant compromises and the free space loss of a communication link becomes the greatest single driver of data bandwidth.

In nearly every case, the bandwidth from Earth to the in-space receiver (the so called forward link or uplink) is smaller by orders of magnitude than the space-to-Earth path (the so called return link or downlink). This is often due to the designer making system trades based on the available ground stations, spacecraft SWAP factors, and the mission itself. Since most space missions are designed to collect large volumes of scientific and other data, and to return them to users on Earth, designers work their system trades to optimize the return bandwidth. Designers of space missions do not have the same luxury of available power, mass, and physical size that terrestrial network designers do. It is just not possible to simply add more processing or network appliances to increase throughput. Rather a delicate balance must be made between mission objectives, instrumentation, spacecraft life, and communications.

17.5 Similarity to Emerging Challenged Terrestrial Networks

The SSI concept is one of several attempts to build networking layers in challenging networking environments. While the SSI concept is, by definition, applicable to space systems its constraints, features, and enabling technologies are similar to a variety of emerging terrestrial networking efforts. Solutions that enable building the SSI can be applied to all of these terrestrial networks in their construction.

17.5.1 Remote Sensing Through IoT Devices

IoT refers to the concept of a large number of small devices working together through a network backbone to accomplish some function. IoT applications can be as mundane as managing the temperature or light levels in a house to as vital as monitoring and managing the flow of electricity through a national power grid. Remote sensing through IoT devices typically involves placing individual devices into one or more sensing networks that, depending on what is being sensed, may not have regular access to power or backhaul networks to egress their data.

Sensor networks experience similar challenged networking environments as nodes on the SSI. Individual sensor nodes may have intermittent connectivity as a function of the terrain and overall geographic footprint where they are deployed (Figure 17.9). Delays in these networks are not necessarily caused by long signal propagation delays, but by frequent disruptions (causing retransmissions) because of low transmit power, the need to wait to accumulate power for any transmission, or waiting for some data mule to pass by near enough to support a transmission. Similarly, remotely deployed sensors may not be able to communicate at high rates as a function of their transmit power.

17.5.2 Wildlife Tracking

In 2003, biology researchers in Kenya were faced with a challenge. They had a requirement to track animals (in this case zebra) over long time periods and long distances and to observe the fine interaction between members of the animals' group, interaction between those animals and others, and what impacts human development has over time. The challenge they faced was that the researchers could only be in the field for a limited time and only periodically. In terms of network design, the problem represented the challenges of mobility (and therefore variable link connectivity), severe SWAP challenges for the instrumentation, and only periodic access to the network (when a researcher was present).

Researchers at Princeton University developed the concept of the "ZebraNet" [17] to address these challenges. Princeton developed a set of custom data acquisition units in the form of collars that could be attached to animals in the field. Each collar contained a set of sensors (GPS, ambient light, accelerometer, etc.), a data logger, and a DTN processor. Sets of collars formed an ad hoc, peer-to-peer network capable of communicating data, status, and control messaging among the group. A master node was also developed to be mounted on a vehicle (ground or

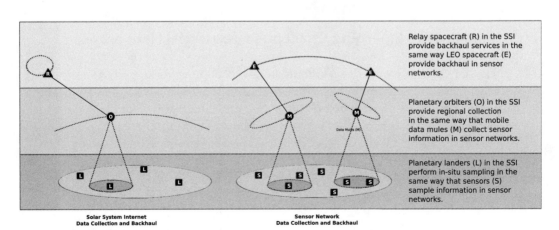

Figure 17.9 Sensor networks are architecturally like the SSI.

aircraft) that would periodically approach the herd, connect to the network, and offload the collected data. There was no real time connectivity between the zebra collars and the broader terrestrial internet. Rather, the master node served as a data mule to collect information in the field and then transfer it to users on the internet upon returning to base.

The concept of instrumenting animals in the field with sensors, storage memory, and a communication system capable of exchanging information on an ad-hoc basis between nodes has grown in popularity in recent years. In the cases of interest, the sensor packages are designed to exchange information with each other in a store and forward manner until a node eventually reaches range of a base station allowing it to dump the entire network's collected information.

17.5.3 Vehicle Data Logging

Vehicle manufacturers have an interest in collecting information on how their cars, boats, and aircraft are performing. Designers use these data to determine how their products are used and how their customers interact with them. Manufacturers can also use these data to provide monitoring and maintenance services as well. Collecting health and status information allows them to recommend preventative maintenance before something breaks and provide diagnostic information when a failure does occur. For example, large aircraft manufacturers, such as Boeing and Airbus, routinely monitor performance telemetry of their aircraft in the field. This information allows them to work with airlines to manage maintenance schedules as well as providing real-world insight into the environments and usage their aircraft are encountering.

The environment in which vehicle data is logged and reported shares several similarities with the SSI. In both cases node mobility causes constant topological change in the network. Often, vehicle data cannot be offloaded from a platform until the vehicle enters some local area such as a garage. Mobility causes limited connectivity — even in cases where vehicles have regular access to some network the amount of engineering data that can be produced may far exceed the bandwidth allocated for health and safety traffic.

17.5.4 Search and Rescue Networks

One of the defining characteristics of disaster recovery efforts is a severe lack of preplaced infrastructure for power, communication, and other public services. In the communication and networking arena this presents as an extreme challenge to establish and maintain connectivity between (often) distant teams of first responders and recovery workers. Since communication and control of a situation is a vital element to quickly reaching people in need of help and to efficient use of limited resources, this challenge is one of the greatest facing disaster recovery efforts.

There are many solutions available to network designers to meet the unique challenges of supporting disaster recovery efforts. Satellite and other persistent communications capabilities can be provided or prepositioned as part of preparation for potential disasters (say along coastal areas in preparation for hurricanes or in the western US in preparation for forest fires.) Such systems can provide highly capable communications capabilities but are dependent on available power (that may be knocked out). They also must survive the initial disaster event. Commercial and government satellite services can provide communications to almost any location on Earth, but how do you ensure that the bandwidth will be available when and where you need it without paying to reserve the service when you aren't using it? Mobile infrastructure can and is brought in with recovery and aid efforts and this typically provides a much-needed link to supporting government, aid service, and medical organizations as well as providing the ability to coordinate large scale logistic of moving supplies into a disaster area and victims out to safety. In fact, disaster recovery command centers are often established at the location of the most reliable communications link specifically for this reason. These mobile communication systems are well suited for a command post, but do not address the needs of the teams of responders, firefighters, and medical personnel who are out in the field looking for victims and stabilizing the situation.

17.6 Features Enabled by Application Patterns

This section discusses the ways in which the patterns discussed in the book can be applied to the SSI networking concept. While certain patterns are meant to be solve different types of problems in a challenged networking environment, other patterns are complimentary (see Table 17.2).

17.6.1 Autonomous Network Management

Deep-space spacecraft operate for extended periods of time without regular bidirectional contact with mission operations centers on Earth. The reasons for these periods of limited connectivity are varied, including oversubscription of groundbases assets, planetary occultations, and individual spacecraft pointing and configuration. The SSI architecture increases the opportunities to communicate with spacecraft by increasing the opportunities for communications paths amongst all assets in the internetwork. However, the solar system will not be densely populated with spacecraft and the SSI will continue to have some bottlenecks between otherwise well-connected constituent networks. When a series of spacecraft and other deployed assets do not have regular contact with Earth, they must implement autonomous operations that preserve their health, their science, and their ability to contribute to the network itself.

Spacecraft employ a variety of well-characterized techniques for in situ, automated fault management such as flying redundant hardware and selecting between them in the event of hardware failure. For complex failures, spacecraft typically

Table 17.2 Delay-Tolerant Applications Enable SSI Features

SSI Networking Features	Offshore Oracle	Training Wheels	Stow-Away	Network Watchdog	Data Forge	Ticket to Ride
Autonomous Network Management	X	X		X		
Inter-Domain Communications			X	X		X
Coordinated Data Fusion			X	X	X	

define a small set of safe operational modes where a spacecraft can be placed in a minimal operating state most likely to preserve both its ability to contact Earth and to prevent any further damage to the vehicle itself. These techniques are effective for providing for the physical health of the spacecraft, but they are not enough to protect the ability of the spacecraft to continue functioning as a networking node in the SSI architecture.

Were SSI nodes to transition to a minimally functional safe mode they would likely lose the ability to transfer the majority of their network traffic. Because safe mode demotions are not planned events, this would take some time to be detected by network operators (particularly if the event happened at a time with little or no connectivity back to Earth). Therefore, SSI nodes must support an autonomous network management function that can preserve network function wherever possible to maintain data flows as part of the SSI.

The patterns of offshore oracle, training wheels, and network watchdog can work together to provide an effect autonomous network management function for local networks in the SSI. Consider the case of multiple planetary orbiters providing a networking backhaul service through a space relay for a series of planetary landers. If one of the orbiters experiences a significant fault such that it must be removed from the local network, some set of autonomy must be in place to detect and handle this event.

In this case, three patterns can be used to track the overall health of the topology. The OOP can be used to collect information from individual spacecraft relating to the overall health of the local network in a readily accessible place. The TWP can be used at all nodes in the SSI to implement monitor-response based reactions. Finally, the NWP can be used to determine when expected data events have failed to occur between nodes in the network.

One possible walkthrough of such a scenario is given in Figure 17.10, which enumerates a series of chronologically ordered events described as follows.

1. Nodes in the local network serve as data sources for an OO located at a boundary between the local network and the rest of the SSI, in this case a space-based relay node. Information is periodically sent to the OO about the state and configuration of the network.
2. At some point, an orbiter may experience a fault that will remove its ability to participate as a functioning node in the internetwork. Local autonomy such as that compliant with the TWP can be used to detect and handle this event for the orbiter.
3. Local watchdogs on nodes expecting to communicate with the orbiter may time out waiting for networked traffic. Local autonomy can be used to adjust to the use of alternate egress points, such as different orbiter passes and the construction of NACKs for the NWP.
4. NACKS, as received by nodes implementing the network watchdog react by updating the topology of the network and sending out configuration updates to remove the lost networking node from communications planning.
5. Query agents on nodes in the local network receive updated topology information from the OO and reconfigure their services around the lost node.

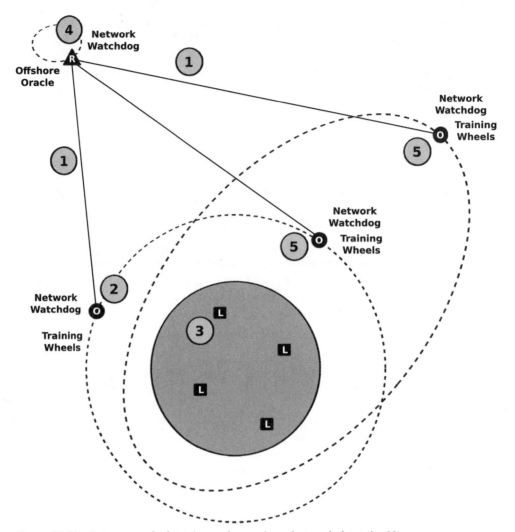

Figure 17.10 Autonomously detecting and removing a lost node from the SSI.

17.6.2 Inter-Domain Communications

As in internetwork, the SSI architecture relies on federating multiple assets provided by all the world's space agencies. In this way, the cost of constructing and maintaining the internetwork is distributed across multiple organizations. Additionally, the cost savings to individual missions naturally leads to the ability to construct more spacecraft that naturally add nodes to the network. As the SSI concept is built out over the next several decades, end-to-end data flows will pass through spacecraft belonging to differing space agencies. Each of these spacecraft will have their own administrative policies and command/control interfaces.

As messages pass through differing administrative domains they may be treated with different priorities and with different monitoring transparency. Developing a trust model for communicating across multiple administrative domains is important because of the consequences of mishandling user data. Relay spacecraft must trust the users whose data they accept—unlike the terrestrial internet where forwarding decisions are made within milliseconds a relay spacecraft may be asked to carry large data volumes for users for extended periods of time. Similarly, data producing nodes must trust their relays to carry data to their destination. Long signal propagation delays and limited storage may require spacecraft to remove data after it has been placed into the network but before it has reached its destination so as to clear space for new observations.

Three patterns naturally work together in this scenario to help navigate interdomain communications. The SAP allows messages to be annotated with a variety of information at various hops in the SSI to ensure that metadata necessary for proper message handling is kept with the user data itself. The TRP uses the concept of stowaways to carry network-specific handling information. Finally, the NWP can be used to provide some indications of message failure if certain domains within the SSI do not provide transparency into message health (see Figure 17.11).

Consider the scenario where a planetary lander owned by one space agency (for example, JAXA) sends a volume of science data to a planetary orbiter operated by a different space agency (for example, ESA), which then relays the data back to Earth via a space relay operated by a third space agency (for example, NASA). Figure 17.11 illustrates this scenario and the steps taken for the data exchange, as explained below.

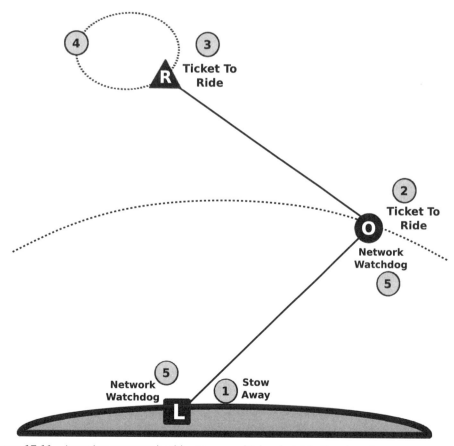

Figure 17.11 Assessing message healthy across multiple administrative domains.

1. The planetary lander produces a series of data observations and packetizes them using an appropriate protocol, such as the BP, which allows for the addition of multiple secondary headers. Some of these headers provide naming, metadata, and security for the messages whose data may only be properly looked at by the lander owner.
2. Once received by a planetary orbiter, the message is authenticated as coming from an appropriate data source and a ticket-to-ride over the orbiter is issued. This ticket describes how the handling of the message should be over-ridden while it exists in the administrative control of the orbiter.
3. At some point, the message is passed to a relay spacecraft. The relay spacecraft authenticates the message as coming from an appropriate source (the orbiter) and issues its own ticket to ride to describe how the message should be handled while in its own administrative domain.
4. At some point, the relay spacecraft may need to drop the message or otherwise fail to deliver it within its time to live. The relay spacecraft may or may not provide an indicator of message failure.
5. A network watchdog running on the orbiter can time out at which point the message can be resent from the orbiter (if the orbiter kept a copy) or a NACK can be generated and sent back to the lander for its own watchdog timer pattern processing. Alternatively, the lander itself can maintain a watchdog and auto-NACK itself based on a failure to receive a confirmation that the message was delivered through the relay.

17.6.3 Coordinated Data Fusion

The benefits of federating multiple spacecraft in the SSI goes beyond adding path diversity to the network. Different spacecraft may be outfitted with a variety of instruments and sensors such that they can provide both their own individual science and the opportunity for coordinated science. Just as a space agency may be able to reduce the cost of an orbiter or a lander by presupposing an existing communications infrastructure, the payloads of the asset may also be simplified if there is already a sensing infrastructure as well.

Coordinated data fusion occurs when multiple payloads across multiple spacecraft generate their own data that can be fused at natural collection points to create new data products. In addition to creating new data products, in situ data analysis can also be used to command when spacecraft should take more data—such as when reacting to new events.

Consider an architecture where multiple landers communicate their sensed data information to a series of planetary orbiters who also relay information back to Earth via a deep-space relay. In this example, orbiters collect and fuse information from landers and may take their own observations as a result of information received from those landers. Figure 17.12 illustrates a series of events, described below, which provide one method of coordinated data fusion.

1. Sensors on planetary landers provide in situ measurements of some planetary surface. These landers may have the same sensor suite or may sense different phenomena. In cases where multiple landers have the same types of sensors, they may generate duplicative data over an overlapping coverage

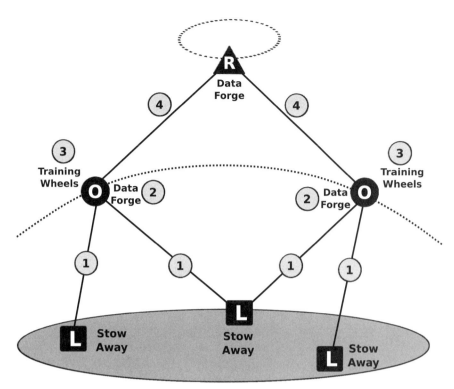

Figure 17.12 Data fusion among spacecraft in the SSI.

areas. Individual landers may not have the ability to coordinate amongst themselves, but they annotate their data and send it to planetary orbiters for storage and egress back to Earth.

2. Orbiters receive sets of data from landers and store them waiting for the opportunity to transmit them via a deep-space relay back to Earth. While waiting for that egress, they may inspect data to remove duplicates and otherwise reduce the data volume as outlined in the DFP.

3. As part of analyzing data returns from landers, orbiters may determine, via their own TWP, that they should also take an observation. For example, if seismic sensors on a planetary lander indicate significant activity, an orbiter may wish to also acquire an overhead view of the area where the lander has been deployed.

4. Information stored at orbiters is relayed back to the deep space relay where it is stored waiting for egress back to Earth. Since different orbiters may have collected similar (or duplicate) information from different (or the same) landers, the DFP is also applied at this layer of the SSI hierarchy to reduce the overall data volume back to Earth.

17.7 Summary

The SSI is a bold concept to apply a networking layer to both near-Earth and deep-space spacecraft. This concept evolved from decades of experience and experimentation relating to the multiple benefits associated with infrastructure reuse. A communications infrastructure around planets and through dedicated space relays allows individual spacecraft to not have to maintain direct-to-Earth connections and provide for the transmit power the require. Further, by allows spacecraft to coordinate their communications, sensing networks can be built to take advantage of other in situ measurements from other spacecraft. This allows new spacecraft to only carry their own unique sensing requirements. Together, a networking concept can significantly reduce the cost of new spacecraft missions, which increases the number of spacecraft that we can practically deploy in our solar system.

The SSI operates in a networking environment challenged by intermittent connectivity, long signal propagation delays, and limitations on bandwidth through any one physical link. These challenges are not significantly different from the types of challenges faced in other emerging terrestrial networking environments. IoT sensor networks, wildlife tracking, vehicle logging, and search-and-rescue operations all seek to benefit from shared communications and sensing infrastructures. While the domains are different, a solution to the SSI will also serve as a solution to these terrestrial network challenges. As the SSI pursues DTN as its enabling set of networking technologies, other terrestrial network deployments are also looking to do the same.

As previously mentioned, a DTN does not solve the problem of making applications function — even when messaging is reliable and delay-tolerant, the applications that govern the management and use of the network must, themselves, also tolerate delays. Capabilities such as autonomous network management, inter-domain communications, and coordinated data fusion can only happen within the SSI in a delay-tolerant way because of the nature of the network itself. To that end, we present example of how delay-tolerant patterns can be used together to provide some of these SSIs unique and motivating capabilities.

17.8 Problems

17.1 Components of the SSI are listed to include ground stations, interplanetary spacecraft, planetary orbiters and landers, and dedicated relay spacecraft. Are there any types of assets missing from this architecture? Explain why or why not.

17.2 Consider the case where a Mars lander communicated both DTE and back to Earth via an orbiter. In what cases would a DTE link be preferred and in what cases would a relay be preferred?

17.3 Assume that you have been asked to design a remote teleoperations experiment similar to the METERON experiments. In your scenario, multiple landed rovers must be teleoperated by astronauts in two different planetary orbiters. Which of the patterns covered in this book could apply to the applications built to provide this capability?

17.4 List another emerging, terrestrial challenged networking environment whose networking problems could be solved by adopting the networking solutions of the SSI. Explain why these environments share the same chal-

17.5 Consider the architecture proposed in Section 17.6.1. Provide another use case other than topology management that would also be satisfied by applying these same delay-tolerant patterns.

17.6 Consider the problem of using tickets-to-ride for interdomain communications and the example scenario presented in Section 17.6.2. Provide three examples of a policy override that may happen when a message is accepted by a new network domain.

17.7 Consider the problem of coordinated data fusion as presented in Section 17.6.3. Provide another example of how lander information could be used to initiate sensors on overhead orbiters and explain how delay-tolerant patterns could help with their implementation.

17.8 Define another useful capability enabled by the SSI and describe how at least two delay-tolerant patterns can be used to guide the design and implementation of applications.

References

[1] *Operations Concept for a Solar System Internetwork (SSI)*, IOAG.T.RC.001.V1. Washington, DC, IOAG, October 15, 2010.

[2] "Report Concerning Space Data System Standards" *(Green Book), Solar System Internetwork (SSI) Architecture*, No. 1, CCSDS 730.1-G-1, Washington, DC, July 2014.

[3] "From Mars to Earth: Getting the Data from the Mars Exploration Rovers to Earth," Space Cameras–Malin Space Science Systems, Malin Space Science Systems, 2005, www.msss.com/mars_images/mars_relay/mars2earth/mer2earth.html.

[4] "Report Concerning Space Data System Standards" *(Green Book), Proximity-1 Space Link Protocol – Rationale, Architecture, and Scenarios*, No. 2, CCSDS 210.0-G-2, Washington, DC, December 2013.

[5] "Recommended Standard" *(Blue Book), Proximity-1 Space Link Protocol—Data Link Layer*, No. 5, CCSDS 211-0-B-5, Washington, DC, December 2013.

[6] "Recommended Standard" *(Blue Book), Proximity-1 Space Link Protocol—Physical Layer*, No. 4, CCSDS 211-1-B-4, Washington, DC, December 2013.

[7] "Recommended Standard" *(Blue Book), Proximity-1 Space Link Protocol–Coding and Synchronization Sublayer*, No. 2, CCSDS 211-2-B-2, Washington, DC, December 2013.

[8] Cerf, V., et. al., "First Deep Space Node on the Interplanetary Internet: The Deep Impact Networking Experiment (DINET)," *Ground System Architectures Workshop*, 2009.

[9] Jones, R., S. Burleigh, J. Torgerson, and J. Wyatt, *Deep Space Networking Experiments on the EPOXI Spacecraft*, 2011, 10.2514/6.2011-1644.

[10] Burleigh, S., "Deep Impact Network Experiment (DINET)," Jet Propulsion Laboratory, California Institute of Technology, Pasadena, CA, March 11, 2008.

[11] Schoolcraft, J., Burleigh, S., Jones, R., Wyatt, J., and Torgerson, L., "The Deep Impact Network Experiments–Concept, Motivation and Results," *SpaceOps 2010 Conference*, American Institute of Aeronautics and Astronautics, https://doi.org/10.2514/6.2010-2262

[12] Schiele, A., "METERON-Validating Orbit-to-Ground Telerobotics Operations Technologies," *11th Symposium on Advanced Space Technologies for Robotics and Automation (ASTRA)*, 2011.

[13] Lii, N. Y., et. al., "Toward Scalable Intuitive Teleoperation of Robots for Space Deployment with the METERON SUPVIS Justin Experiment," *Proc. of the 14th Symposium on Advanced Space Technologies for Robotics and Automation (ASTRA)*, 2017.

[14] Gifford, K., et. al., "BioNet Middleware and Software Framework in Support of Space Operations," *SpaceOps 2010 Conference, Delivering on the Dream*, Hosted by NASA Marshall Space Flight Center and Organized by AIAA, 2010.

[15] Schlesinger, A., et. al., "Delay/Disruption Tolerant Networking for the International Space Station (ISS)," *2017 IEEE Aerospace Conference*, 2017, pp. 1–14.

[16] Basciano, T., B. Pohlchuck, and N. M. Shamburger, "Application of Delay Tolerant Networking on the International Space Station," *NASA Technical Reports*, 2019.

[17] Zhang, P., et. al., "Hardware Design Experiences in ZebraNet," *SenSys 2004*, ACM, pp. 227–238.

About the Authors

Edward J. Birrane is a computer scientist and software engineer who focuses on the adaptation of computer networking protocols and software for use in nontraditional environments, such as space communications. He has supported a variety of flight software engineering efforts, such as those for the NASA New Horizons mission to explore Pluto and the Parker Solar Probe mission to explore the Sun's corona. Dr. Birrane currently works with industry, government, and academia on the design and development of protocols for delay-/disruption-tolerant networking (DTN). Within the Internet Engineering Task Force (IETF) and the Consultative Committee for Space Data Systems (CCSDS), he is the author or coauthor of the Bundle Protocol v7, the Bundle Protocol Security Protocol (BPSec), and the Asynchronous Management Protocol (AMP).

Dr. Birrane is a member of the principal professional staff at the Johns Hopkins University Applied Physics Laboratory where he manages the Embedded Applications Group within their Space Exploration Sector. He is an adjunct professor of computer science at both the University of Maryland, Baltimore County, and the Johns Hopkins University. He lives in Maryland with his wife, Linda, and two children.

Jason Soloff is a systems engineer with over 17 years of service to NASA robotic and human space flight programs, including positions at NASA's Goddard Space Flight Center, Johnson Space Center, and Headquarters. Mr. Soloff's experience includes communications systems engineering for near-Earth and deep-space missions, leading avionics and communications for the Constellation Program, providing systems architectures for human exploration and operations, developing communications and networking interoperability approaches and policies among the international space community, and serving as NASA's Chief Engineer for Space Systems Protection. Mr. Soloff also served among NASA's representatives to the IOAG, SCAWG, and CCSDS. He is the cofounder and chief technology officer of farSight Technologies, LLC, developing advanced communication, networking, PNT, and sensing technologies for the scientific, defense, ISR, and aerospace communities.

Index

A

Absolute overhead, 203, 205–6
Abstract base class (ABC), 153–54
Access points
    adding, mobility issues and, 20
    approach to populating, 18–19
    mobile, 20–21
    multiple, nodes traversing, 19
Acknowledged serialization, 229, 230
Addressing
    DTNs, 99
    local, 43
    naming and, 99, 151
    overlay networks, 134–39
Ad hoc networking, 89
ALOHA, 162
Annotative headers, 260
Application Agent (AA), 102
Application annotations, 118
Application behavior, terrestrial internet, 36
Application design patterns (ADPs)
    about, 152–53
    abstract base class (ABC) and, 153–54
    applying to existing software, 155–56
    defined, 152
    documentation format, 156–58
    as generalized solution, 156
    history and concept of, 153–54
    object-oriented analysis and design (OOAD) principles and, 153–54
    poor solution representation, 155
    sample, 155
    sample architectural smells, 156
    value of, 154–56
Application developers, 7–8
Application endpoints, 162
Application layer challenges
    growing data volumes, 25–26
    growing delays, 24–25
    increased disruptions, 25
    See also Challenged networking environments
Application services
    about, 145
    core networking services, 149–50, 170
    dependencies, 147
    endpoint services, 151–52
    federated services, 150–51, 170
    hierarchy, 145–52
    illustrated elements, 146
    overlay services, 151, 170
    three-tiered model, 146
    transport services, 148–49, 170
    types of, 147
Architectural smells, 156
ARQ schemes, 163
Asymmetry, 77–78
Audience, this book, 7–8
Automatic computing, 47–50
Autonomous network management, 300–302

B

Bandwidth, 50, 273, 296–97
Bidirectionality, 76–77
BioServe Commercial Generic Bioprocessing Apparatus (CGBA), 294
Bit error rate (BER), 22–23
Block confidentiality blocks (BCBs), 214–15
Block integrity blocks (BIBs), 214–15
BP agents (BPAs)
    behavior of, 117

BP agents (BPAs) (continued)
   bundle status reporting mechanism, 235
   as component of DTN ecosystem, 100
   content caching and, 121
   defined, 4, 100
   extensions, dissimilar, 277
   responsibilities, 101
   status report generation, 174
BP ecosystem
   about, 99–100
   Application Agent (AA), 102
   BP Agent (BPA), 100–101
   convergence layer adapters (CLA), 100
   convergence layers, 100
BP-enabled concepts, 118–21
BP Security Specification (BPSec), 213–14
Broadcast messaging, 135–37
Bundle Protocol (BP)
   additional processing, 124–25
   administrative records, 234–36
   application annotations and, 118
   bundle structure, 113–14
   content caching and, 121
   custody transfer and, 120
   defined, 4
   DTN support illustration, 99
   extension blocks, 116–18, 119, 123–24
   fragmentation, 123
   goals, 109
   implemented at various layers, 112
   mapping of status reports, 176
   naming and addressing, 151
   optimal fragment size, 123
   payload block, 116
   PDU, 113
   primary block, 114–16
   protocol layering considerations, 111–13
   security, 122
   services supported, 98
   services unique to, 111
   special considerations, 121–25
   state of standard, 98
   status reason codes, 175
   status reporting, 174–76
   status report reason codes, 235
   storage management, 121–22
   store and forward, 110–11, 204
   summary, 126
   versions of, 66, 112
   virtual path, 161–62

C

Carrier-sense multiple access (CSMA), 138
Case studies
   data forage pattern (DFP), 251–53
   network watchdog pattern (NWP), 234–36
   Noah's Data Ark, 280–81
   offshore oracle pattern (OOP), 174–77
   Solar System Internet (SSI), 287–306
   stow away pattern (SAP), 213–16
   ticket-to-ride pattern (TRP), 268–70
   training wheels pattern (TWP), 195–97
Certificate Authorities (CAs), 39
Challenged networking environments
   defined, 13
   error handling in, 26
   frequent/long-lived impairments, 27
   link layer, 16–21
   mapping of information to messaging in, 241
   network layer, 22–24
   network resource preservation in, 40
   separating responsibilities, 14
   summary, 30
   terrestrial internet approaches to, 36–43
Circular dependencies, 194
Code smell, 156
Common message protocol structures, 202
Concept of operations (CONOPs), 7
Configuration sources (CSs), 167–68
Connectivity
   concept of operations (CONOPs), 28
   high, maintaining, 28
   Solar System Internet (SSI), 75, 296
   temporal, 139
   Wi-Fi, 35
Consultative Committee for Space Data Systems (CCSDS), 64–65, 66, 150, 276
Contact Graph Routing (CGR), 215–16
Content caching
   application designs for, 152
   as BP-enabled concept, 121

Index 313

in challenged networks, 161–78
Content delivery networks (CDNs)
  autonomic applications requirement, 48
  content caches, 44–45
  defined, 43
  as overlays, 43
  pattern examples, 44
Control and data flows
  data forage pattern (DFP), 245–47
  network watchdog pattern (NWP), 229
  offshore oracle pattern (OOP), 168–69
  stow away pattern (SAP), 208
  ticket-to-ride pattern (TRP), 264–65
  training wheels pattern (TWP), 188–90
Control flows
  DFP, 246
  TRP, 264
Control plane
  implementation of, 184
  importance of, 184
  information, changing, 267
  local, 185–89
  remote, 185
  Convergence layer adapters (CLA), 100
  Convergence layers
    defined, 100
    many-to-many, 105
    many-to-one, 104–5
    multiple, 104–5
    one-to-many, 105
Coordinated data fusion, 304–5
Core Flight System (CFS), 195–97
Core networking services, 149–50, 170
Custody transfer, 120
Cyclic redundancy check type, 115

D

DARPA (Defense Advanced Research Projects Agency), 56–58, 66, 79
Data derivation engine (DDE), 185, 192
Data dissemination methods, 39–40, 41
Data flows
  NWP, 229
  OOP, 168–69
  SAP, 210
  TWP, 189

Data forage pattern (DFP)
  case studies, 251–53
  control and data flows, 245–47
  control flows through, 246
  data expiration and, 248
  data fusion engine (DFE), 243, 244, 245, 249, 250
  defined, 239
  in information-centric networking (ICN), 252–53
  information destinations (IDs), 243, 244, 245, 247
  information sources (ISs), 243, 244, 246
  integration, 247–49
  local data cache (LDC), 243, 244, 245–47, 249, 250
  mechanism, 241
  pattern context, 239–42
  pattern overview, 243–47
  pattern use, 247–48
  problem being solved, 242–43
  pros and cons, 250–51
  recommended design decisions, 248–49
  role definitions, 244–45
  semantic data validator (SDV), 243, 244, 245–47
  service types, 247
  in store-and-forward routing applications, 251–52
  summary, 253
  system components, 244
  what can go wrong, 250
Data fusion, coordinated, 304–5
Data fusion engine (DFE), 243, 244, 245, 249, 250
Data identification, 187
Data mules, 89
Data placement strategies, 38–39
Data planes, 184
Data pruning and fusing, 242
Data repositories (DRs), 185, 187, 188–89
Data subscriptions, 45–47
Data volumes, growing, 25–26
Data workflows
  defined, 34
  terrestrial internet, 34–35
Deep Impact Networking (DINET)

DINET I, 291–92
DINET II, 292, 293
   experiments, 290–92
   scenario, 291
Deep-space instruments
   distributed, 82–83
   formation flight, 83–84
   MAXIM, 82–83
Deep-space spacecraft, 48
Delay-/disruption-tolerant research, 55–56
Delays
   common, 24
   growing, 24–25
   as problematic, 24–25
Delay tolerance
   DTN protocol and, 8
   efficiency and, 5
Delay-tolerant applications, 4–5
Delay-tolerant architectures (DTAs), 96, 97
Delay-tolerant networks (DTNs)
   accepting reactive fragments, 275
   as architectural approach, 4
   architectures for, 96
   BP ecosystem, 99–102
   capabilities, advent of, 9
   capabilities, assumptions, and constraints of, 5–6
   concept, 3
   data pruning and fusing in, 242
   desirable properties, 97
   dissimilar implementations, 276–80
   DTNS as not guaranteeing, 6
   end-to-end communication support, 4
   example data flow, 6
   heterogenous networks, 275–76
   implementation of, 4
   initial motivator for research, 69
   in the IRTF, 65
   motivations for, 93
   naming and addressing, 99
   operational deployment of, 1
   as overlay network, 78
   packets, 5
   protocols, 98–99
   research into, 55
   resource limitations, 273–75
   special node characteristics, 102–5

   time as an input and, 166
   in violating network assumptions, 165
   waypoints, 5
   what can go wrong, 273–83
Design patterns. See Application design patterns (ADPs)
Disruptions
   endpoints and, 165
   increased, 25
   of optical links, 88
Dissimilar implementations
   about, 276–77
   BPA extensions, 277
   operating environments, 278–80
   service level expectations, 277–78
Distributed and mobile sensor webs, 86–87
Distributed spacecraft constellations
   about, 79–80
   deep-space instruments, 82–86
   planetary observation missions, 80–81
   in-space communication and navigation, 84–86
   space-ground integration, 80–81
Domain name server (DNS), 38, 39
DTN Research Group (DTNRG), 57, 65
Dykstra's algorithm, 251

E

Earth-to-Mars network, 176, 177
Efficiency
   delay tolerance and, 5
   metric, 2
   protocol, increasing, 29–30
Encapsulating interfaces, 133
Encapsulation, 262, 266
Encapsulation view, 14–15
Endpoints
   application, 162
   coordinating between, 164
   coordination, delays and disruptions and, 165
   physical link, 161–62
   protocol layers leading to multiple types of, 162
   references, message overhead and, 203–4
   virtual path, 165

End-to-end data loss, small, 95
Error reporting, 261
European Data Relay Satellite System (EDRS), 88
Extensible payload system (EPS), 206–7, 208–9, 210
Extension blocks
    BPAs and, 117
    defined, 116
    handling, 123–24
    previous node block (PNB), 117–18
    to summarize payload data, 119
    utility, 117
Extension cache (EC), 206, 207–8, 211
Extension policy engine (EPE), 263, 264–65, 267

F

False negative, 223
False positive, 223
Federated deep-space networking, 269–70
Federated internetworks, 140–41
Federated services
    accounting for differences and, 150
    defined, 150
    end-to-end network, 151
    OOs, 170
    See also Application services
Feedback, timely and reliable, 94
Field-programmable gate arrays (FPGAs), 124
Fire-and-forget networking operations, 162
5G, 3
Formation flight, 83–84
Forwarding, 149–50
4G LTE, 2–3, 21
Fragmentation, BP, 123

G

Goodput, 2, 205

H

Heterogeneity, link, 21
Heterogenous networks, 275–76
Hypertext Transfer Protocol (HTTP), 50

I

IEEE 802.11, 17

If This, Then That (IFTTT) automation, 48–49
Information-centric networking (ICN), 252–53
Information destinations (IDs), 243, 244, 245, 247
Information placement, 37–39
Information sources (ISs), 243, 244, 246
Integration
    data forage pattern (DFP), 247–49
    network watchdog pattern (NWP), 231–33
    offshore oracle pattern (OOP), 171–73
    stow away pattern (SAP), 209–12
    ticket-to-ride pattern (TRP), 265–67
    training wheels pattern (TWP), 192–94
Inter-domain communications, 302–4
International space agencies
Interoperability Plenary (IOP), 58, 59, 64–65
IOAG and, 59–60, 65
    Space Communications Architecture Working Group (SCAWG), 60–64
    Space Internetworking Strategy Group (SISG), 64
International Space Station, 294–95
Internet Research Task Force (IRTF)
    defined, 57
    DTN in, 65
Interplanetary Internet Research Group (IPNRG), 65
Interoperability Plenary (IOP), 58, 64–65
Interplanetary Internet Research Group (IPNRG), 65
IOAG, establishment of, 59–60
IP messaging, 138–39

L

Late binding, 103–4
Latency
    metric, 2
    Solar System Internet (SSI), 76
Layering view, 14–15
Lazy evaluation approach, 189–90, 191
Lickliter Transmission Protocol (LTP), 164
Link characteristics, altering, 28–29

Link heterogeneity, 21
Link layer challenges
　high-rate wireless communication, 16–18
　link heterogeneity, 21
　node mobility, 18–21
　　See also Challenged networking environments
Local addressing, 43
Local control planes (LCPs), 185, 188, 189
Local data cache (LDC), 243, 244, 245–47, 249, 250
Local data collectors (LDCs), 185, 186, 188–89
Local node watchdogs (LNWs), 225, 231, 232
Local watched components (LWCs), 225, 226, 229–31, 232
Long Term Evolution (LTE), 16

M

Macro environment, 279–80
Managed locality controller (MLC), 225, 226, 228, 229–31, 233–34
Managed locality watchdog (MLW), 225, 226–27, 231, 233
Many-to-many, 105
Many-to-one, 104–5
Mapping of information to messaging, 241
Marketing metrics, 1
Mars relay communications, 289–90
Message fragmentation, 212
Micro-Arcsecond X-ray Imaging Mission (MAXIM), 82–83, 84
Micro environment, 278–79
MIL-STD-1553, 150
Mission operators, 8
Mobile access points, 20–21
Multicast messaging, 137–38
Multiple convergence layers, 104–5
Multi-Purpose End-To-End Robotic Operation Network (METERON), 292–94

N

Naming and addressing, 99, 151
NASA
　Advanced Exploration Projects (AES), 66
　Advanced Exploration Systems and International Space Station Programs, 294–95
　DARPA funded activities, 57
　Deep Space Network, 70
　DSN Ground Stations, 70, 71
　Goddard Space Flight Center, 57, 74, 79
　issues and challenges, 58
　JPL, 56, 73, 79, 86
　Orbiting Carbon Observatory 2 (OCO-2), 81
　SensorWeb Network, 86, 87
　Space Network Fleet, 72
　Space Network Geosynchronous Relay System, 72
　Tracking and Data Relay Satellite (TDRS), 71, 72
　A-Train and C-Train, 81
　Wallops Flight Facility, 70
　　See also Space Communications Architecture Working Group (SCAWG)
Near-field communication (NFC), 21
Negative acknowledgement (NACK) schemes, 220, 222
Network address translation (NAT), 282
Network architects, 7
Network boundaries, 257–59
Network error conditions
　approaches to handling, 27–30
　defined, 26–27
　link characteristics, altering, 28–29
　nodes, increasing number and, 27–28
　protocol efficiency, increasing, 29–30
Network faults, 231
Networking architectures
　about, 129
　federated internetworks, 140–41
　implementation of, 158
　overlay networks, 131–39
　partitioned networks, 139–40
　summary, 141
Networking services
　configuring each other, 149
　core, 149–50
　OOP, 170
Network layer challenges
　path losses, 22–23

time-variant, partitioning topologies, 23
unsynchronized node information, 23–24
    See also Challenged networking environments
Network policies, terrestrial internet, 33–34
Network pricing and economics, terrestrial internet, 35–36
Network segment extension (NSE), 263–65, 266–67
Network service layers
    defining, 15–16
    high fidelity decomposition of, 15
    three-layer model, 15
Network watchdog pattern (NWP)
    acknowledged serialization, 229, 230
    in BP administrative records, 234–36
    case studies, 234–36
    control and data flows, 229
    data flows through, 229
    defined, 219
    integration, 231–33
    local node watchdogs (LNWs), 225, 231, 232
    local watched components (LWCs), 225, 226, 229–31, 232
    managed locality controller (MLC), 225, 226, 228, 229–31, 233–34
    managed locality watchdog (MLW), 225, 226–27, 231, 233
    negative acknowledgement (NACK) schemes, 220, 222
    network communications and, 232
    pattern context, 219–22
    pattern overview, 224–29
    pattern use, 231–32
    positive acknowledgement (ACK) schemes, 220, 222, 223
    problem being solved, 223–24
    processing bundle status reports and, 236
    pros and cons, 234
    recommended design decisions, 232–33
    role definitions, 226–27
    service types, 229–31
    summary, 236–37
    system components, 227
    time-tagged commanding, 228, 230
    what can go wrong, 233–34
Noah's Data Ark, 280–81
Node mobility
    adding access points and, 20
    in disrupting wireless links, 19
    mobile access points and, 20
    nub-and-spoke networking architecture and, 18, 19
    populating access points and, 18–19
Nodes, increasing number of, 27–28
Nub-and-spoke networking architecture, 18, 19

O

Object-oriented analysis and design (OOAD) principles, 153
Offshore oracle pattern (OOP)
    augmentation, 166
    in BP status reporting, 174–76
    case studies, 174–77
    component control flows, 171
    configuration sources (CSs), 167–68
    control and data flows, 168–69
    data flows through, 169
    defined, 164
    integration, 171–73
    pattern context, 161–64
    pattern overview, 166–69
    problem being solved, 165–66
    pros and cons, 174, 175
    Query Applications (QAs), 168, 169, 172
    recommended design decisions, 172–73
    reporting sources (RSs), 166–68
    role definitions, 166–68
    in security policy updates, 177
    service types, 170–71
    summary, 178
    in topology management, 176–77
    what can go wrong, 173
    when to use, 171–72
One-to-many, 105
Open Systems Interconnection (OSI) model, 62
Operating environments
    dissimilar, 278–80
    macro, 279–80

Operating environments (continued)
  micro, 278–79
Optical communication, 87–89
Optical links, disruption and handoff of, 88
Organization, this book, 8–9
Overhead measurements, 203
Overlay networks
  about, 131–32
  broadcast messaging, 13–17
  defined, 131
  illustrated, 281–82
  as insulated from encapsulation, 134
  IP messaging, 138–39
  as logical subsets of physical networks, 132
  in motion, 282
  multicast messaging, 137–38
  network addressing schemes, 134–39
  pass-through, 131–32
  routing with space names, 281–82
  unicast messaging, 134–35, 136
  working with, 281–83
  See also Networking architectures
Overlay services, 151, 170
Oversampling, 242–43, 248

P

Packet error rate (PER), 22–23
Packets, DTN, 5
Partitioned networks, 139–40
Pass-through interfaces, 133
Path existence, 94
Path losses, 22–23
Patience, 4
Pattern context
  data forage pattern (DFP), 239–42
  network watchdog pattern (NWP), 219–22
  offshore oracle pattern (OOP), 161–64
  stow away pattern (SAP), 201–4
  ticket-to-ride pattern (TRP), 257–59
  training wheels pattern (TWP), 181–83
Pattern overview
  data forage pattern (DFP), 243–47
  network watchdog pattern (NWP), 224–29

offshore oracle pattern (OOP), 166–69
stow away pattern (SAP), 206–8
ticket-to-ride pattern (TRP), 262–65
training wheels pattern (TWP), 185–90
Pattern use
  data forage pattern (DFP), 247–48
  network watchdog pattern (NWP), 231–32
  offshore oracle pattern (OOP), 171–72
  stow away pattern (SAP), 209–11
  ticket-to-ride pattern (TRP), 266
  training wheels pattern (TWP), 192–93
Payload block, 116
Payload extension processor (PEP), 206, 207–8, 212
Performance abstraction, 95–96
Persistent storage, 103
Physical boundaries, 258, 259
Physical link endpoints, 161–62
Place information, 37–39
Planetary observation missions, 80–81
Policy boundaries, 258, 259
Positive acknowledgement (ACK) schemes, 220, 222, 223
Predicate evaluator (PE), 185, 187–88, 189
Previous node blocks (PNBs), 116–17
Primary block
  CRC field, 116
  creation timestamp, 115
  cyclic redundancy check type, 115
  destination EID, 115
  fragment offset, 116
  lifetime, 116
  processing control flags, 114–15
  report-to EID, 115
  source node ID, 115
  versions, 114
  See also Bundle Protocol (BP)
Problem being solved
  data forage pattern (DFP), 242–43
  network watchdog pattern (NWP), 223–24
  offshore oracle pattern (OOP), 165–66
  stow away pattern (SAP), 204–6
  ticket-to-ride pattern (TRP), 259–62
  training wheels pattern (TWP), 184–85

Processing capacity, 274
Processing control flags, 114–15
Processing flags, 260–61
Processing time, 195
Pros and cons
    data forage pattern (DFP), 250–51
    network watchdog pattern (NWP), 234
    offshore oracle pattern (OOP), 174
    stow away pattern (SAP), 213
    ticket-to-ride pattern (TRP), 268
    training wheels pattern (TWP), 195, 196
Protocol boundaries, 258, 259
Protocol efficiency, 29–30
Publish-subscribe (PubSub) pattern
    benefits of using, 47
    centralized and distributed models, 45
    defined, 45
Pull methods, 39, 41
Push data, 39–40, 41

Q

Query Applications (QAs), 168, 172
Quick UDP Internet Connections (QUIC), 50–51

R

Rate and range, 18
Reactive fragments, accepting, 275
Recommended design decisions
    data forage pattern (DFP), 248–49
    network watchdog pattern (NWP), 232–33
    offshore oracle pattern (OOP), 172–73
    stow away pattern (SAP), 211–12
    ticket-to-ride pattern (TRP), 266–67
    training wheels pattern (TWP), 193–94
Relative overhead, 203
Reliability, 13
Remote data aggregators (RDAs), 185, 187, 188
Reporting sources (RSs), 166–68
Representational State Transfer (REST), 50
Resource limitations, 273–75
Resource utilization, 261–62
RESTful interfaces, 50, 51
Retransmission penalties, 213
Retransmissions, 205
RFC4838, 93–94
RF link performance, 17, 22
Role definitions
    data forage pattern (DFP), 244–45
    network watchdog pattern (NWP), 226–27
    offshore oracle pattern (OOP), 166–68
    stow away pattern (SAP), 106–7
    ticket-to-ride pattern (TRP), 263–64
    training wheels pattern (TWP), 186–88
Round trip time (RTT), 76
Routing loops, 212

S

Search and rescue networks, 299–300
Security, BP, 122
Security association identifier (SAID), 203
Security certificates, 183
Security policy updates, 177
Segment extension editor (SEE), 263, 264–65, 267
Semantic data validator (SDV), 243, 244, 245–47
Service level expectations, dissimilar, 277–78
Service types
    data forage pattern (DFP), 247
    network watchdog pattern (NWP), 229–31
    offshore oracle pattern (OOP), 170–71
    stow away pattern (SAP), 208–9
    ticket-to-ride pattern (TRP), 265
    training wheels pattern (TWP), 190–92
Sessionless data exchange, 41–43
Sessions
    avoiding, 40–43
    defined, 40–41
    lifecycle illustration, 42
    phases, 41
    TCP, 163
Simple Network Management Protocol (SNMP), 48–49
Solar System Internet (SSI)
    architecture for, 78–79
    artist depiction, 80
    asymmetry, 77–78

Solar System Internet (SSI) (continued)
    autonomous network management, 300–302
    bandwidth, 296–97
    bidirectionality, 76–77
    challenges, 295–97
    connectivity, 75, 296
    coordinated data fusion, 304–5
    Deep Impact Networking (DINET) experiments, 290–92
    defined, 287
    delay, 296
    example propagation delays, 297
    experiences and experiments, 289–95
    features enabled by application patterns, 300–305
    idea proposal, 74
    inter-domain communications, 302–4
    International Space Station, 294–95
    latency, 76
    Mars relay communications, 289–90
    METERON, 292–94
    motivations for, 288–89
    overview, 287
    remote sensing through IoT devices, 297–98
    search and rescue networks and, 299–300
    similarity to emerging challenged terrestrial networks, 297–300
    summary, 305–6
    vehicle data logging and, 299
    wildlife tracking and, 298–99
Space Communications Architecture Working Group (SCAWG)
    charter, 60
    defined, 60
    Final Report recommendation, 60–61
    information and data flow architecture models, 62
    layered communications architecture, 63
    services management model, 63
    services model, 62
    services model comparison, 64
    in-space communication and navigation, 85–86
    space communication architecture, 61–62

Spacecraft fault protection
    Core Flight System (CFS) and, 196–97
    defined, 195
    example assumptions, 196
Space-ground integration, 80–81
Space Internetworking Strategy Group (SISG), 64
Space links, 77–78
Special node characteristics
    about, 102–3
    late binding, 103–4
    multiple convergence layers, 104–5
    persistent storage, 103
Specific, measurable, achievable, relevant, and timely (SMART), 183
Stale mitigations, 195
Standard model for networking, 130–31
Stateless data, 50–51
Storage capacity, 274
Storage management, 121–22
Store and forward, BP, 110–11, 204
Store-and-forward routing applications, 251–52
Stow away pattern (SAP)
    in adding message-related/message-agnostic information, 212
    in BP security extensions, 213–15
    case studies, 213–16
    in CGR extensions, 215–16
    control and data flows, 208
    data flows through, 210
    defined, 201
    extensible payload system (EPS), 206–7, 208–9, 210
    extension cache (EC), 206, 207–8, 211
    integration, 209–12
    message fragmentation, 212
    pattern context, 201–4
    pattern overview, 206–8
    pattern use, 209–11
    payload extension mechanism, 209
    payload extension processor (PEP), 206, 207–8, 212
    problem being solved, 204–6
    pros and cons, 213
    recommended design decisions, 211–12

Index 321

retransmission penalties, 213
role definitions, 106–7
routing loops, 212
service types, 208–9
summary, 216
system roles of, 207
what can go wrong, 212–13
Subscription models, 45–47
System engineers, 7

T

TCP/IP networking, 148
TCP/IP ubiquity, 95
Telemonitoring (TM) protocol, 150
Temperature sensing, 243
Term endpoint identifier (EID), 99
Terrestrial internet
   application behavior, 36
   approaches to challenged networking environments, 36–43
   assumptions made by, 93–96
   automatic computing, 47–50
   challenges in, 33–36
   content delivery networks (CDNs), 43–45
   data subscriptions, 45–47
   data volumes, growing, 25–26
   data workflows, 34–35
   design patterns, 43–51
   network policies and, 33–34
   network pricing and economics, 35–36
   path existence, 94
   performance abstraction, 95–96
   place information, 37–39
   push data, 39–40
   sessionless data exchange, 40–43
   small end-to-end data loss, 95
   stateless data, 50–51
   summary, 51–52
   TCP/IP ubiquity, 95
   timely, reliable, and actionable feedback, 94
Terrestrial Planet Finder (TPF) missions, 82, 83–84
Ticket-to-ride pattern (TRP)
   adding to network segment boundary, 260

   case studies, 268–70
   component overview, 263
   control and data flows, 264–65
   control flow through, 264
   defined, 257
   extension policy engine (EPE), 263, 264–65, 267
   in federated deep-space networking, 269–70
   integration, 265–67
   network boundaries and, 257–59
   network segment extension (NSE), 263–65, 266–67
   pattern context, 257–59
   pattern overview, 262–65
   pattern use, 266
   potential conflict examples and, 260–62
   problem being solved, 259–62
   pros and cons, 268
   recommended design decisions, 266–67
   role definitions, 263–64
   segment extension editor (SEE), 263, 264–65, 267
   service types, 265
   summary, 270–71
   what can go wrong, 267–68
Timeliness, 13
Time-tagged commanding, 228, 230
Time-variant, partitioning topologies, 23
Topology convergence, 130–31
Topology management, 176–77
TPF-1/Darwin, 84
TPF-O concept, 85
Tracking and Data Relay Satellite (TDRS), 71, 72
Training wheels pattern (TWP)
   assumptions, 181–83, 192–93
   case study, 195–97
   circular dependencies, 194
   control and data flows, 188–90
   control loop, 189
   data derivation engine (DDE), 185, 192
   data flow illustration, 189
   data repositories (DRs), 185, 187, 188–89
   data timeliness, 194

Training wheels pattern (TWP) (continued)
   defined, 181
   dynamic assumptions, 182–83
   integration, 192–94
   lazy evaluation approach, 189–90, 191
   local control planes (LCPs), 185, 188, 189
   local data collectors (LDCs), 185, 186, 188–89
   mitigations to failed assumptions, 193
   pattern context, 181–83
   pattern overview, 185–90
   pattern use, 192–93
   predicate evaluation, 193–94
   predicate evaluator (PE), 185, 187–88, 189
   predicate expressions as independent, 193
   problem being solved, 184–85
   processing time, 195
   proper function, 192
   pros and cons, 195, 196
   recommended design decisions, 193–94
   remote data aggregators (RDAs), 185, 187, 188
   role definitions, 186–88
   sampled and derived data, 194
   service types, 190–92
   in spacecraft fault protection, 195–97
   stale mitigations, 195
   static assumptions, 182
   summary, 197–98
   system roles of, 186
   timing data collection, 190
   what can go wrong, 194–95
Transport protocols, 266
Transport services, 148–49, 170
Tripartite model, 240

U

Unicast messaging, 134–35, 136

Universal Resource Identifier (URI), 99
Unsynchronized node information, 23–24
Use cases
   ad hoc network and data mules, 89
   distributed and mobile sensor webs, 86–87
   distributed spacecraft constellations, 79–86
   motivating, 69
   optical communication, 87–89
   Solar System Internet (SSI), 69–79
   summary, 89
   value of, 69

V

Vehicle data logging, 299
Virtual BP path, 161–62
Virtual path endpoints, 165

W

Waypoints, DTN, 5
What can go wrong
   data forage pattern (DFP), 250
   network watchdog pattern (NWP), 233–34
   offshore oracle pattern (OOP), 173–74
   special considerations for DTNs, 273–83
   stow away pattern (SAP), 212–13
   ticket-to-ride pattern (TRP), 267–68
   training wheels pattern (TWP), 194–95
Wi-Fi connectivity, 35
Wildlife tracking, 298–99
Wireless communication
   high-rate, link layer challenge, 16–18
   state of, 1–3

Z

ZebraNet, 280

Artech House
Space Technology and Applications Series

5G and Satellite Spectrum, Standards, and Scale, Geoff Varrall

Business Strategies for Satellite Systems, D. K. Sachdev

Delay-Tolerant Satellite Networks, Juan A. Fraire, Jorge M. Finochietto, and Scott C. Burleigh

Designing Delay-Tolerant Applications for Store-and-Forward Networks, Edward J. Birrane, Jason A. Soloff

Filter Design for Satellite Communications: Helical Resonator Technology, Efstratios Doumanis, George Goussetis, and Savvas Kosmopoulos

Gigahertz and Terahertz Technologies for Broadband Communications, Terry Edwards

Ground Segment and Earth Station Handbook, Bruce R. Elbert

Introduction to Satellite Communication, Third Edition, Bruce R. Elbert

Laser Space Communications, David G. Aviv

Low Earth Orbital Satellites for Personal Communication Networks, Abbas Jamalipour

Mobile Satellite Communications, Shingo Ohmori, et al.

Radio Frequency Inteference in Communications Systems, Bruce R. Elbert

Radio Interferometry and Satellite Tracking, Seiichiro Kawase

Satellite Broadcast Systems Engineering, Jorge Matos Gómez

The Satellite Communication Applications Handbook, *Second Edition*, Bruce R. Elbert

Satellite Communications Fundamentals, Jules E. Kadish and Thomas W. R. East

Satellite Communications Network Design and Analysis, Kenneth Y. Jo

The Satellite Communication Ground Segment and Earth Station Handbook, Second Edition, Bruce Elbert

Space Microelectronics, Volume 1, Modern Spacecraft Classification, Failure, and Electrical Component Requirements, Anatoly Belous, Vitali Saladukha, and Siarhei Shvedau

Space Microelectronics, Volume 2, Integrated Circuit Design for Space Applications, Anatoly Belous, Vitali Saladukha, and Siarhei Shvedau

Understanding GPS: Principles and Applications, Second Edition, Elliott D. Kaplan and Christopher J. Hegarty, editors

For further information on these and other Artech House titles, including out-of-print books available through our In-Print-Forever® (IPF®) program, contact:

Artech House
685 Canton Street
Norwood, MA 02062
Phone: 781-769-9750
Fax: 781-769-6334
e-mail: artech@artechhouse.com

Artech House
16 Sussex Street
London SW1V 4RW U.K.
Phone: +44 (0)171-973-8077
Fax: +44 (0)171-630-0166
e-mail: artech-uk@artechhouse.com

Find us on the World Wide Web at: www.artechhouse.com